2021北京国际首饰艺术展

2021 BEIJING INTERNATIONAL JEWELLERY ART EXHIBITION

主 编｜詹炳宏　　　　副主编｜兰翠芹｜高伟｜胡俊
EDITOR | ZHAN BINGHONG　　DEPUTY EDITORS | LAN CUIQIN | GAO WEI | HU JUN

中国纺织出版社有限公司

内 容 提 要

2021北京国际首饰艺术展以"砺行"为主题,旨在展现不同文化的交流与碰撞,推进多元首饰艺术创作潮流,并为不同的艺术创作观念和思潮搭建互动的平台与空间。本次展览共收到全球33个国家和地区共748位艺术家的作品,最终评选出300余位艺术家参展。本书精心挑选其中具有代表性的艺术作品,力求展现展览的引领性、多样性以及启发性。

本书图文并茂,图例丰富,适合高等院校珠宝首饰专业师生、珠宝首饰设计师、收藏家以及广大首饰爱好者阅读与参考。

图书在版编目(CIP)数据

2021北京国际首饰艺术展 / 詹炳宏主编;兰翠芹,高伟,胡俊副主编 . — 北京:中国纺织出版社有限公司,2023.4

ISBN 978-7-5229-0302-6

Ⅰ . ① 2⋯ Ⅱ . ①詹⋯ ②兰⋯ ③高⋯ ④胡⋯ Ⅲ . ① 首饰 — 设计 — 作品集 — 世界 — 现代 Ⅳ . ① TS934.3

中国国家版本馆 CIP 数据核字 (2023) 第 021420 号

责任编辑:李春奕　　特约编辑:徐铭爽
责任校对:高函　　责任印制:王艳丽

中国纺织出版社有限公司出版发行
地址:北京市朝阳区百子湾东里 A407 号楼　邮政编码:100124
销售电话:010—67004422　传真:010—87155801
http://www.c-textilep.com
中国纺织出版社天猫旗舰店
官方微博 http://weibo.com/2119887771
北京华联印刷有限公司印刷　各地新华书店经销
2023 年 4 月第 1 版第 1 次印刷
开本:787 × 1092　1/8　印张:36
字数:95 千字　定价:468.00 元

Honing

2021 北京国际首饰艺术展·砺行

2021 BEIJING INTERNATIONAL
JEWELLERY ART EXHIBITION

2021 北京国际首饰艺术作品展

展览汇聚了来自全球 33 个国家和地区的 225 位艺术家、设计师的作品。其中包括英国中央圣马丁艺术学院荣休教授 Carlion Broadhead、意大利当代首饰艺术学会主席 Maria Rosa Franzin 等国际著名首饰艺术家。

北京国际首饰艺术展开幕式及参展作品秀

区别于国际首饰艺术静态展，本次展览以动态时尚走秀的形式呈现世界各国艺术家及设计师的首饰作品。

未来域首饰设计大赛

未来域首饰设计大赛的目的旨在弘扬珠宝首饰理论，推动创新发展，聚合来自创新、技术、智造文化等多层次多维度的设计指向，让传统与时尚、传承与创新在这里同频共振，深耕珠宝行业的潜力，积极探索时尚与未来生活的平衡，使优秀的设计获得更多的社会关注，通过设计师的演绎，开创我国珠宝首饰行业新时代的辉煌。肩负使命，向新而行。

全球首饰设计教育峰会

针对首饰艺术教育话题，探讨国际语境下的首饰艺术，商业首饰的发展方向、趋势和未来首饰设计人才的发展需求。

第五届北京国际首饰艺术展高峰论坛

该论坛汇聚了具有影响力的首饰专家学者、产业精英和珠宝企业家，是在中国境内举办的最高水平的国际首饰艺术与设计学术活动。

2021 Beijing International Jewellery Art Exhibition

The exhibition brings together the works of 225 artists and designers from 33 countries and regions around the world. Among them are carlion broadhead, Professor Emeritus of the Central Saint Martin School of Art in the UK; Maria Rosa Franzin, President of the Italian Contemporary Jewellery Art Society, and other world renowned Jewellery artists.

Opening Ceremony of Beijing International Jewellery Art Exhibition and Exhibition Works

Different from the other international jewellery art static exhibition, the show presents the jewellery works of artists and designers from all over the world in the form of dynamic fashion shows.

Future Field Jewellery Design Competition

The purpose of the Future Field Jewellery Design Competition is to carry forward Jewellery theory, promote innovation and development, gather multi-level and multi-dimensional design directions from innovation, technology, intelligent manufacturing culture, etc; let tradition and fashion, inheritance and innovation resonate at the same frequency here, deeply cultivate the potential of the jewellery industry, actively explore the balance between fashion and future life, and find excellent design to get more social attention. Through the deduction of designers to create a new brilliante era in China's jewellery industry. Shoulder the mission and move towards the new era.

Global Jewellery Design Education Summit

Aiming at the topic of jewellery art education, this meeting discusses the development direction and trend of jewellery art and commercial jewellery in the international context, as well as the development demand of jewellery design talents in the future.

The 5th Summit Forum of Beijing International Jewellery Art Exhibition

The forum brings together influential jewellery experts and scholars, industry elites and jewellery entrepreneurs. It is the highest level jewellery art and design international academic activity held in China.

本次展览的主题为"砺行"。《山海经·西山经》有云：西南三百六十里，曰崦嵫之山，其上多丹木，其叶如楮，其实大如瓜，赤符而黑理，食之以瘅，可以御火。其阳多龟，其阴多玉。苕水出焉，而西流注于海，其中多砥砺。"砥"指细腻的磨刀石，"砺"指粗糙的磨刀石，为磨炼、锻炼、激励、勉励之意。"行"，人之步趋也，本义为"路"，引申为"践行"。主题"砺行"，意指疫情之下，全球的政治、经济与文化艺术生活都受到了前所未有的冲击，作为人类命运共同体的一员，无论前路多么逶迤，我们都应当望向远方，携手并肩，砥砺前行。

This year's exhibition is themed "Li Xing". As the traditional Chinese myth *Shanhaijing·Xishanjing* describes: There is a mountain called Yanzi, which is one hundred and eighty kilometres away on the south-western part. There are lots of Dan Mu on Yanzi Mountain. The leaves of Dan Mu are like paper mulberry leaves. The fruit of Dan Mu is as big as watermelon. The sepal of Dan Mu is red with dark lines. The fruit of Dan Mu are able to cure malaria, and has the function of resisting internal heat. There are lots of turtles in the south of Yanzi, and lots of jade in the north of Yanzi. The Shao River originates from Yanzi, then it goes westward into the sea. In the Shao River, there are many grindstones, named "Di Li" in ancient Chinese. "Di" refers to the fine grindstone and "Li" refers to the rough grindstone. "Di Li" means to temper, train, encourage and exhort. In Chinese, "Xing" means the trend of a man's step. Its original meaning is road, which is then extended to mean "practice". Above all, the theme "Li Xing" means under the influence of pandemic, our economic, politic, culture and art life have all been subject to unprecedented impact. As a member of the community of shared future for mankind, no one can detach himself or herself from this interconnected world. Winding or straight, the road is always in front of us. With our eyes on the future, we should walk hand in hand and encourage each other to forge ahead.

组织方式
2021 北京国际首饰艺术展

指导单位：

教育部：中外人文交流中心
工信部：工业文化发展中心

主办单位：

北京服装学院

承办单位：

服饰艺术与工程学院
艺术设计学院
国际首饰设计高校联盟
中国生活方式设计研究院
中关村时尚产业创新园（BIFT PARK）

协办单位：

北京设计学会
《艺术设计研究》
《设计》

学术支持：

《艺术设计研究》《设计》
《中国宝石》《中国珠宝首饰》
《中国黄金珠宝》《芭莎珠宝》及 Klimt02

支持单位：

外国院校： 英国皇家艺术学院、英国伦敦艺术大学、英国布莱顿大学、英国曼彻斯特都市大学、英国伯明翰城市大学、意大利佛罗伦萨阿契米亚首饰学院、意大利佛罗伦萨欧纳菲首饰学院、意大利帕多瓦塞瓦蒂克国立艺术学院、比利时安特卫普皇家美术学院、德国慕尼黑美术学院、德国普福尔茨海姆应用技术大学、西班牙巴塞罗那玛萨纳艺术学院、美国罗德岛设计学院

中国院校： 清华大学美术学院、中央美术学院、中国地质大学、北京工业大学艺术设计学院、天津美术学院、中国美术学院、上海大学美术学院、南京艺术学院、山东工艺美术学院、鲁迅美术学院

企业公司： 北京菜市口百货股份有限公司、北京珐琅厂有限责任公司、北京金一文化发展股份有限公司、北京玉尊源玉雕艺术有限责任公司、沈阳萃华金银珠宝股份有限公司、国金有限公司、上海老凤祥有限公司、上海豫园黄金珠宝集团有限公司、上海铂利德钻石有限公司、上海华泰珠宝商场有限公司、上海张铁军珠宝集团有限公司、南京通灵珠宝股份有限公司、周大福珠宝金行有限公司、周生生珠宝金行有限公司、周大生珠宝股份有限公司、广东潮宏基实业股份有限公司、深圳市百泰珠宝首饰有限公司、深圳市粤豪珠宝有限公司、深圳市甘露珠宝首饰有限公司（爱得康）、深圳市盛峰黄金有限公司、深圳市沃尔弗斯实业有限公司、中和盛世珠宝文化发展有限公司

ORGANIZATION MODE

2021 BEIJING INTERNATIONAL JEWELLERY ART EXHIBITION

Guided by:

Ministry of Education (MOE): China Center for International People-to-People Exchange
Ministry of Industry and Information Technology (MIIT): Industrial Culture Development Center

Sponsored by:

Beijing Institute of Fashion Technology

Organized by:

School of Fashion Accessory Art & Engineering
School of Art & Design
International Jewellery Colleges Association
China Academy of Lifestyle Design
Beijing Institute of Fashion Technology Park (BIFT PARK), a fashion industry innovation park in Zhongguancun, Beijing

Co-organized by:

Beijing Design Society
Art & Design Research
Design

Academic Support:

Art & Design Research, Design,
China Gems, China Jewellery,
China Gold & Jewellery,
Barzar Jewellery, Klimt02

Supporting Organizations:

Foreign Colleges and Universities: Royal College of Art (the UK), University of the Arts London (the UK), University of Brighton (the UK), Manchester Metropolitan University (the UK), Birmingham City University (the UK), Alchimia Contemporary Jewellery School (Florence, Italy), Le Arti Orafe Jewellery School & Academy (Florence, Italy), Pietro Selvatico Art School (Padua, Italy), Royal Academy of Fine Arts (Antwerp, Belgium), Academy of Fine Arts (Munich, Germany), Pforzheim University of Applied Sciences (Germany), La Massana Fine Arts School (Barcelona, Spain), Rhode Island School of Design (the USA), Academy of Arts & Design

China's Colleges and Universities: Tsinghua University (Academy of Arts & Design), Central Academy of Fine Arts, China University of Geosciences, Beijing University of Technology (College of Art and Design), Tianjin Academy of Fine Arts, China Academy of Art, Shanghai Academy of Fine Arts, Nanjing University of the Arts, Shandong University of Arts & Design, Lu Xun Academy of Fine Arts

Enterprise Companies: Beijing Caishikou Department Store Co., Ltd., Beijing Enamel Factory Co., Ltd., Beijing Kingee Culture Development Co., Ltd., Beijing Yuzunyuan Jade Art Co., Ltd., Shenyang Cuihua Gold and Silver Jewellery Co., Ltd., Guojin Co., Ltd., Shanghai Laofengxiang Co., Ltd., Shanghai Yuyuan Jewellery Group Co., Ltd., Shanghai Bolide Diamond Co., Ltd., Shanghai Huatai Jewellery Mall Co., Ltd., Shanghai Zhangtiejun Jewellery Group Co., Ltd., Leysen Jewellery Inc., Chow Tai Fook Jewellery Group Limited, Chow Sang Jewellery Co., Ltd., Chow Tai Seng Jewellery Co., Ltd., Guangdong CHJ Industry Co., Ltd., Shenzhen Baitai Jewellery Co., Ltd., Shenzhen Yuehao Jewellery Co., Ltd., Shenzhen Ganlu Jewellery Co., Ltd. (Aidekang), Shenzhen Shengfeng Gold Co., Ltd., Shenzhen Wolfers Industry Co., Ltd. and Zhonghe Shengshi Jewellery Culture Development Co., Ltd

组织委员会

2021 北京国际首饰艺术展

主席：

贾荣林（北京服装学院校长）

副主席：

杜柯伟（教育部中外人文交流中心主任）
罗民（工信部工业文化发展中心主任）

秘书长：

詹炳宏（北京服装学院副校长）

组委会办公室：

主任： 兰翠芹、张德权
副主任： 李雪梅、孙钊、王碧涛、姚文琴
成员： 高伟、张弘、胡俊、邹宁馨、傅永和、赵祎、潘峰、宋懿、熊芏芏、唐天、程之璐、刘小奇、王涛、韩欣然、王浩睿

委员：

白静宜、包晓莹、陈火龙、储卫民、才大颖、程学林、陈国珍、陈晓华、常炜、杜半、范海民、韩雨蒙、郭强、郭颖、郭英杰、郭新、高伟、洪兴宇、胡书刚、黄雯、Laurent-Max De Cock（比利时）、Leo Caballero（西班牙）、兰翠芹、李春珂、李峻、李秀美、李英杰、李雪梅、刘骁、廖创宾、马世忠、Maria Rosa Franzin（意大利）、任进、Peter Deckers（新西兰）、潘团结、齐红、宋处岭、施堃、施健、孙仲鸣、唐绪祥、滕菲、王春利、王春刚、王晓昕、王志伟、汪正虹、文乾刚、许平、许梦佳、月文、张纯辉、张福文、张铁成、张世忠、张卫峰、赵丹绮、郑静、郑裕彤、郑耿坚、钟连盛、周桃林、周厚厚、周宗文、朱伟明、庄冬冬、邹宁馨

执行委员：

高伟、胡俊、傅永和、赵祎、潘峰、宋懿、熊芏芏、唐天、程之璐、刘小奇、王涛、韩欣然、王浩睿

ORGANIZING COMMITTEE

President:

Jia Ronglin, President of Beijing Institute of Fashion Technology (BIFT)

Sponsored by:

Du Kewei, Director of China Center for International People-to-People Exchange, MOE Luo Min, Director of Industrial Culture Development Center, MIIT

Secretary General:

Zhan Binghong, Vice President of Beijing Institute of Fashion Technology (BIFT)

Organizing Committee Office:

Directors: Lan Cuiqin, Zhang Dequan
Vice Directors: Li Xuemei, Sun Zhao, Wang Bitao and Yao Wenqin
Members: Gao Wei, Zhang Hong, Hu Jun, Zou Ningxin, Fu Yonghe, Zhao Yi, Pan Feng, Song Yi, Xiong Dudu, Tang Tian, Cheng Zhilu, Liu Xiaoqi, Wang Tao, Han Xinran and Wang Haorui

Committee Members:

Bai Jingyi, Bao Xiaoying, Chen Huolong, Chu Weimin, Cai Daying, Cheng Xuelin, Chen Guozhen, Chen Xiaohua, Chang Wei, Du Ban, Fan Haimin, Han Yumeng, Guo Qiang, Guo Ying, Guo Yingjie, Guo Xin, Gao Wei, Hong Xingyu, Hu Shugang, Huang Xiaowang, Huang Wen, Laurent-Max De Cock (Belgium), Leo Caballero (Spain), Lan Cuiqin, Li Chunke, Li Jun, Li Xiumei, Li Yingjie, Li Xuemei, Liu Xiao, Liao Chuangbin, Ma Shizhong, Maria Rosa Franzin (Italy), Ren Jin, Peter Deckers (New Zealand), Pan Tuanjie, Qi Hong, Song Chuling, Shi Kun, Shi Jian, Sun Zhongming, Tang Xuxiang, Teng Fei, Wang Chunli, Wang Chungang, Wang Xiaoxin, Wang Zhiwei, Wang Zhenghong, Wen Qiangang, Xu Ping, Xu Mengjia, Yue Wen, Zhang Chunhui, Zhang Fuwen, Zhang Tiecheng, Zhang Shizhong, Zhang Weifeng, Zhao Danqi, Zheng Jing, Zheng Yutong, Zheng Gengjian, Zhong Liansheng, Zhou Taolin, Zhou Houhou, Zhou Zongwen, Zhu Weiming, Zhuang Dongdong, Zou Ningxin

Executive Committee:

Gao Wei, Hu Jun, Fu Yonghe, Zhao Yi, Pan Feng, Song Yi, Xiong Dudu, Tang Tian, Cheng Zhilu, Liu Xiaoqi, Wang Tao, Han Xinran and Wang Haorui

Honing

2021 北京国际首饰艺术展·砺行

2021 BEIJING INTERNATIONAL
JEWELLERY ART EXHIBITION

砺行
Honing

序言

2021 北京国际首饰艺术展

首饰大概可以追溯到遥远的石器时代，因宗教功能动因和社会功能动因而起源和发展。伴随着人类历史的进程，首饰的原材料、佩戴人群、象征意义和作用都在日渐丰富。时代变迁中首饰历久弥新，向今世人诉往昔事。中外首饰领域的设计师、匠人、从业者们为满足人民日益提升的精神审美和物质需求，不断地求索探讨创新。伴随着生活水平的提高，首饰在人与人之间，在文明之间的沟通、交流中发挥着独特的作用。开展首饰领域的研究交流和创新，有利于珠宝产业的创新发展和相关领域人才的培养，有助于增进不同民族和文化之间的交流理解和欣赏。北京服装学院坚持求实创新，学以致用的办学理念形成了服装引领、艺工融合的办学特色，建设了中关村服饰时尚设计产业创新园。北京服装学院，以独特的办学优势和鲜明的办学特色，富有成效的服装服饰教育实践，对我国服装设计时尚和文化创意人才培养，以及产业发展做出了突出贡献。在中华优秀传统文化的传承和弘扬方面发挥了重要作用。北京服装学院依托办学优势与教学特色，举办北京国际首饰艺术展这个国际学术平台，既体现了人民对美好生活的向往和提高人才培养质量的更高追求，也搭建了促进人文交流的实践平台。

PREFACE

Jewellery can probably be traced back to the remote stone age, which originated and developed due to religious and social function drivers. With the progress of human history, the raw materials, wearers, symbolic significance and functions of jewellery are increasingly enriched. With the changes of the times, jewellery has been renewed over the years, telling the world about the history. Designers, artisans and practitioners in the field of jewellery at home and abroad constantly seek, explore and innovate to meet the people's increasing spiritual aesthetic and material needs. With the improvement of living standards, jewellery plays a unique role in the communication and exchange between people and civilizations. Carrying out research exchange and innovation in the jewellery field is conducive to the innovative development of the jewellery industry and the cultivation of talents in related fields, and helps to enhance the exchange, understanding and appreciation among different nationalities and cultures. Beijing Institute of Fashion Technology adheres to the school running concept of seeking innovation and learning skills for use, forming the school running characteristics of clothing leading the integration of art and industry, and building the BIFT park. With its unique advantages, distinctive school running characteristics and fruitful clothing education practice, Beijing Institute of Fashion Technology has made unique and outstanding contributions to the training of fashion and cultural creative talents in China's clothing design and industrial development. It has played an important role in inheriting and carrying forward the excellent traditional Chinese culture. Beijing Institute of Fashion Technology relies on its advantages and teaching characteristics. The holding of Beijing International Jewellery Art Exhibition, an international academic platform, reflects the higher pursuit of meeting people's yearning for a better life and improving the quality of talent training. It has also built a practical platform to promote people-to-people and cultural exchanges.

北京服装学院连续 10 年，举办了五届北京国际首饰艺术展，致力于打造国际学术平台，促进我国设计学科建设。本届展览的主题为"砺行"，意指疫情之下全球的政治、经济与文化艺术生活，都受到了前所未有的冲击。作为人类命运共同体成员，无论前路多么逶迤，我们都应当望向远方，携手并肩砥砺前行。首饰，通过造物让身体与美对话，构成人类美好生活的重要载体。在历史长河中通过精工细作的劳动教化丰富了人类的生物属性。这些闪耀的璀璨之物又进一步佐证了人类改造自然的能力。

当首饰佩戴在人的身体上，与身体的一动一静、一张一弛交错融合在一起时，便具有了超越静态物体的体验感受，参与到身体动势和衣着搭配的动态变化中，成为彰显人们品位与文化价值的点睛之笔。今天的首饰艺术动态秀，将为大家展示本届首饰展中最为精彩的 151 件艺术作品。希望引导、关注、展开对首饰身份等的定义与思辨，欣赏人对材料的表达，以及对技艺的探索。了解首饰造物中金属的质感是对金属材料的极致拓展和创新利用，体验艺术家幽默的生活态度和戏剧性，洞察首饰背后独特的创意视角。

要增强文化自信，以美为媒，加强国际文化交流。北京国际首饰艺术展作为国际文化交流的平台。通过首饰展示文化自信，传播中国文化，传播中国美，展示中国力量并积极促进不同文化之间的互信交流与合作，是一个创造美、展示美、分享美的国际舞台，这个舞台与美同行，希望一直精彩。

贾荣林

北京服装学院校长

Beijing Institute of Fashion Technology has held five Beijing International Jewellery Art Exhibitions for 10 consecutive years, and it is committed to building an international academic platform and promoting the construction of design disciplines in China.

The theme of this exhibition is "Li Xing", which refers to the global political, economic, cultural and artistic life under the epidemic. As a member of the community of shared future for mankind, no matter how winding the road ahead is, we should look far away and forge ahead hand in hand. Jewellery, through creation, enables the body to talk with beauty, and constitutes an important carrier for a better life for mankind. In the history, the biological attributes of human beings have been enriched through the labor education of exquisite work. These shining things further prove the ability of human beings to transform nature.

When the jewellery is weared on the human body and intertwined with the movement, static, tension and relaxation of the body, it has the experience of going beyond the static object, participates in the dynamic changes of the body momentum and clothing collocation, and becomes the highlight of people's taste and cultural value. Today's jewellery art dynamic show will show you 151 of the most wonderful works of art in the jewellery exhibition. I hope the audience to pay attention to the definition and thinking of jewellery identity, and appreciate people's expression of materials and exploration of skills. Understanding the texture of metal in jewellery creation is the ultimate expansion and innovative use of metal materials, experience the artist's humorous life attitude and drama, and gain insight into the unique creative perspective behind jewellery.

Enhance cultural self-confidence, take the beauty as the medium to strengthen international cultural exchanges. Beijing International Jewellery Art Exhibition serves as a platform for international cultural exchanges. Jewellery shows cultural confidence, spreads Chinese culture, spreads Chinese beauty, shows Chinese power, and actively promotes mutual trust, exchange, and cooperation between different cultures. It is an international stage for creating, displaying, and sharing beauty. This stage goes with beauty, and I hope it will always be wonderful.

Jia Ronglin

Headmaster of
Beijing Institute of Fashion Technology

前言
2021 北京国际首饰艺术展

"披罗衣之璀璨兮，珥瑶碧之华琚。戴金翠之首饰，缀明珠以耀躯。"

——曹植《洛神赋》

首饰源于身体装饰，代表着人类对美的追求，更承载了人的思想、情感和价值观。首饰的起源与人类文明的起源同步。首饰曾象征至高无上的精神和信仰，也曾代表真诚而恒久的爱情与亲情。早在旧石器时代，北京山顶洞人就开始佩戴钻孔串起的石珠、蚌壳、兽骨等饰品；希腊的爱琴文明出土了历史久远的黄金饰品；荷马史诗中对迈锡尼的评价是"rich in gold"。首饰中蕴含着民族、人文、地域、历史、自然、思想、情感等诸多因素。

在多元化发展的今天，首饰不仅仅是提高生活品质的重要艺术媒介，也是文化的载体，首饰设计长期以独特的艺术视角对人的身体与美学表达进行孜孜不倦地探索。今天的造物者从不曾停歇对首饰独特艺术魅力的诠释，更加不惮于探索新的形式与媒介、技艺与手段。首饰设计探索未来科技，精研传承技艺，延展艺术交互关系，未来更加呈现出多元化发展的创新态势。

FOREWORD

"Wrapped in the soft rustle of a silken garments, she decks herself with flowery earrings of jade. Gold and kingfisher hairpins adorning her head, strings of bright pearls to make her body shine."

——*The Ode to the Goddess of the Luo* by Cao Zhi

Jewellery originates from body adornment, which reflects thoughts, emotions and values of human being. The origin of jewellery synchronized with the origin of human civilization. It was used to symbolize the supreme spirit and belief, and represent sincere, lasting romance and family affection. As early as the Paleolithic Age, the Upper Cave Man began to wear jewellery such as stone beads, clam shells and animal bones. Greek Aegean civilization unearthed gold ornaments with a long history. The evaluation of Mycenae in Homer's Epic is "rich in gold". Jewellery contains factors such as nationality, humanity, region, history, nature, thought, emotion and so on.

Nowadays, under the trend of pluralistic development, jewellery is not only an important artistic medium to improve the quality of life, but also a carrier of culture. Jewellery design has been tirelessly exploring human body and aesthetic expression from a unique artistic perspective for a long time. Today's makers never stop interpreting the unique artistic charm of jewellery, and dare to explore new situations, media, skills and means. Jewellery design explores the future science and technology, carefully studies and inherits craftsmanship, and extends the interactive relationship of art, showing an innovative trend of pluralistic development in the future.

本届展览将聚焦国际首饰行业的发展趋势与时尚动态、人工智能与首饰未来技术方向、可持续设计理念下的首饰教育以及全球化背景下开展首饰人才培养路径等。

北京服装学院一直致力于设计创造美好生活，与美同行，引领时尚产业发展趋势。为此，我们打造高水平的国际学术平台。北京国际首饰艺术展是全球首饰艺术与设计的品牌活动，旨在推进多元化、可持续的首饰艺术设计趋势，打造首饰艺术设计全球对话平台。本次活动包括国际首饰艺术动态秀、国际首饰艺术高峰论坛、国际首饰设计教育峰会等内容。期望延续历届首饰艺术展的盛况，探讨国际首饰艺术的最新思想，引领未来发展方向。

预祝 2021 北京国际首饰艺术展圆满成功！

周志军

北京服装学院党委书记

This exhibition focuses on the development and fashion trend of international jewellery industry, the trend of artificial intelligence and jewellery technology in the future, education under the concept of sustainable design, and the way of cultivating jewellery talents under the background of globalization.

Beijing Institute of Fashion Technology (BIFT) is devoted to designing and creating the beautiful life, grows with beauty, and leads the development trend of fashion industry. Therefore, we construct the international academic platform. Beijing International Jewellery Art Exhibition includes a series of brand activities of global jewellery art and design. This exhibition is intended to promote jewellery design trend of diversification and sustainability and set up an interactive platform that embraces the exchange of different ideas from all over the world. During the exhibition, International Jewellery Art Runway Show, International Jewellery Art Summit Forum and Global Jewellery Design Education Summit will be held. It is expected to build on the achievements of previous exhibitions, explores the latest ideas of international jewellery art and leads the future development direction.

Finally, I wish 2021 Beijing International Jewellery Art Exhibition a great success!

Zhou Zhijun

Party Committee Secretary of
Beijing Institute of Fashion Technology

致辞

2021 北京国际首饰艺术展

女士们，先生们，中外人文交流是党和国家对外工作的重要组成部分，是夯实中外关系的社会民意基础，是增强我国软实力、提高对外开放水平的重要内容和途径，是"一带一路"倡议和人类命运共同体建构的基础和支柱。人文交流的核心是人与人的交流，心与心的沟通。所以人类交流的理念，人文交流理念，包含以人为本、开放平等、尊重、包容、理解。北京服装学院已连续举办五届国际学术研讨，今晚将 T 台表演和首饰设计展示相结合，为首饰展示开创了一个更加动态多维的欣赏角度。更蕴含着对不同文化的价值观交流互鉴的追求，借此机会我想与大家分享两点希望：一是通过活动强化人文交流理念，衷心希望以首饰为媒，帮助当代青年学生加深对人文交流理念的理解和认识，培养深厚的人文素养，不断从古今中外文明成果中汲取有益营养。挖掘中华优秀传统文化，传承中华文明，坚定文化自信，成长为德、智、体、美、劳全面发展且具有国际化视野的一代新人。二是通过活动提升人们交流的能力，当今时代是经济全球化的时代，我们也面临着百年未有之大变局，世界进入动荡变革期。衷心地希望通过开展活动，增进中外学生和学者们的交流，拓展国际视野，提升跨文化理解能力和用人文交流理念增进交流促进合作的能力。讲好中国故事，传播好中国声音，为夯实人类命运共同体的民心相通基础作出贡献。

杨小春

教育部中外人文交流中心副主任

SPEECH

Ladies and gentlemen, cultural exchanges between China and foreign countries are an important part of the party's and the country's external work. It is the way of consolidating public support for sino-foreign relations. It is an important content and way to strengthen China's soft power and improve the level of opening up. It is the foundation and pillar of the Belt and Road initiative and the joint construction of community with shared future for mankind. The core of culture exchange is people-to-people exchange and heart-to-heart communication. Therefore, the concept of human exchange, and the concept of cultural exchange, include people-oriented ideas, open, equal, respect, tolerance and understanding. Beijing Institute of Fashion Techolohy has held five consecutive international academic seminars. Tonight, the combination of T-stage performance and Jewellery design exhibition has created a more dynamic and multidimensional appreciation angle for jewellery display. It also contains the pursuit of the exchange and mutual learning of values of different cultures. I want to share two hopes with you by this opportunity. Firstly, strengthen the concept of cultural exchange through activities, and sincerely hope to use jewellery as a medium to help contemporary young students deepen their understanding of the concept of cultural exchange and cultivate profound humanistic quality. We should constantly draw beneficial nutrition from the achievements of civilization at all times and in all countries, excavate the excellent Chinese traditional culture, inherit the Chinese civilization, strengthen cultural confidence, and grow into a new generation with all-round development of morality, intelligence, physique, beauty and labor and an international vision. Secondly, improve people's ability to communicate through activities. Today is the era of economic globalization. We are also facing great changes that have not been seen in a century. The world has entered a period of turbulence and change. We sincerely hope that through the activities, we can enhance the exchange between Chinese and foreign students and scholars, expand international horizons, improve cross-cultural understanding and the ability to enhance exchanges and promote cooperation with the concept of people-to-people exchange. Tell the story of China well, spread the voice of China well, and make contributions to consolidating the foundation of the community of a shared future for mankind.

Yang Xiaochun

Deputy Director of China Center for International People-to-People Exchange, Ministry of Education

IJCA 设计宣言

首饰设计的社会责任行动纲领

首饰承载人的思想与情感，是人类发展历程中人文社会、时代精神、历史积淀的观念创造以及视觉感受。在多元化发展的今天，首饰已成为人类生活的重要艺术媒介和文化载体，超越具有审美意义的"美"，成为人们对美好生活的新期待。大家在关注当代群体的行为方式、价值观、社会环境以及未来发展的基础上，将首饰作为桥梁和媒介，探索社会赋予它的新使命。今天，首饰设计师已经从单纯追求形式上的创新，转向从更深刻的层次探索设计与人类发展的关系，力图通过设计活动，在人、社会、环境之间建立一和和谐发展的机制。在新的时代背景下，首饰的社会功能也有了更为多元的创新表现，反映了全球化时代人们的行为方式和价值构成。今天的造物者从未停歇对首饰独特艺术魅力的诠释，更加不惮于探索新的形势与媒介、技艺与方法。首饰设计全方位介入推动社会文明的发展，它探索科技、精研技艺、延伸交互与关系，呈现出多元化的创新态势。

当下，人类命运交织于同一个共同体语境。在全球化视域下，我们一致倡导发表《IJCA 设计宣言 —— 首饰设计的社会责任行动纲领》。

IJCA DESIGN MANIFESTO
SOCIAL RESPONSIBILITY ACTION PLAN FOR JEWELLERY DESIGN

Jewellery, which carries people's thoughts and emotions, is the concept creation and visual perception of human society, the spirit of the times, and historical accumulation in the history of human development. In the era of diversified development, jewellery has become an important artistic medium and cultural carrier of human life, and it has surpassed the "beauty" with aesthetic significance and become people's new expectations for a beautiful life. Based on the attention paid to the behaviors, values, social environment and future development of contemporary groups, jewellery is used as a bridge and medium to explore the new mission. Currently, jewellery designers have turned from simply pursuing formal innovation to exploring the relationship between design and human development at a more profound level, trying to take advantage of design activities to establish a harmonious development mechanism among people, society and environment. Under the new historical background, the social function of jewellery also has more diversified innovation, reflecting the behavior pattern and value composition of people in the age of globalization. Today's creators have never stopped interpreting the unique artistic charm of jewellery, and they dare to explore new situations, media, techniques, and methods. Jewellery design is completely involved in every aspect to promote the development of social civilization. Besides, it explores science and technology, carefully studies techniques, and extends interaction and relations, thus presenting a diversified innovation situation.

At present, the destiny of human beings is intertwined in the same community context. From the perspective of globalization, we jointly advocate the publication of *The IJCA Design Declaration — Social Responsibility Action Plan for Jewellery Design*.

IJCA 设计宣言

共同倡议

1. 倡导设计师回归社会责任感

引导设计师践行新时代语境下的创新转型，从首饰与社会、首饰与群体关系的基本视角出发，讨论首饰如何立足自身特性进行多维度的社会创新，对社会以及特定群体产生多元化的影响。鼓励首饰设计师回归社会责任感，将首饰视为一种"载体"完成"沟通""传达""导向"的社会使命。将设计师的角色从"造物者"，转化为话题或事件的"发起者"，构建社会沟通的对话机制，让首饰成为连接人与人的关键要素。

2. 树立发展社会责任教育的意愿

将社会责任作为首饰设计教育的核心诉求，引导设计师具备社会化创新意识。随着社会发展的更迭，首饰为了更好地服务社会和大众，其旧有的功能和属性需要革新，必然释放出更多的潜能与空间。这就需要教育者重新思考首饰设计教育的核心使命：如何以首饰为支点鼓励青年设计师承担更多的社会责任，促使设计与社会的有机互动。在微观的教育场景中，鼓励学生积极探索人、环境、物品的深度关系，培育具有社会责任感、全球视野及未来竞争力的设计领导者。

3. 鼓励超越商业逻辑的价值输出

设计的目标并不仅是追求形式多样或是商业利益价值的产出，而是用更为负责任的方式去影响他人，体现设计师角色的社会价值。以社会创新为诉求的设计活动及其所提倡的生活方式、生产方式，可以成功地将个人利益与社会利益、环境利益统一，是推动面向可持续发展的路径之一。在首饰的范畴中，回归社会责任感，把解决社会问题、寻求社会影响力作为设计的开端，可以成为首饰社会创新的起点，从而提出更多的首饰设计解决方案，使之达到商业价值与社会价值的平衡。

我们共同倡议，以社会责任驱动首饰设计创新，融汇全球设计资源和集体智慧，保持开放的交流合作态度，凝聚经验，共享资源，共同推动首饰设计教育事业大发展！

IJCA DESIGN MANIFESTO

1. Advocate Designers to Return to the Sense of Social Responsibility

Guiding designers to practice innovative transformation in the context of the new era, discuss how jewellery can carry out multi-dimensional social innovation based on its own characteristics and have diversified impacts on society and specific groups from the basic perspective of the relationship between jewellery and society, jewellery and groups. Encourage jewellery designers to return to the sense of social responsibility and regard jewellery as a "carrier" to complete the social mission of "communication", "transmission" and "guidance". Transform the role of designers from "creators" to "initiator" of topics or events, construct the dialogue mechanism of social communication, and let jewellery be a key element connecting people.

2. Establish the Willingness to Develop Social Responsibility Education

Take social responsibility as the core appeal of jewellery design education and guide designers to equipe the consciousness of social innovation. With the change of social cevelopment, the old functions and attributes of jewellery need to be innovated in order to better serve society and the public, which will inevitably release more potentials and space. Therefore, this requires educators to rethink the core mission of jewellery design education: how to encourage young designers to assume more social responsibilities by taking jewellery as the supporting point, thus promoting the organic interaction between design and society. In the micro educational scenario, encourage students to actively explore the deep relationship among people, environment and objects, and cultivate design leaders with a sense of social responsibility, global vision and future competitiveness.

3. Encourage Value Output beyond Business Logic

The goal of design is not only to pursue the output of various forms or commercial benefits, but to influence others in a more responsible way and reflect the social value of the designers' role. Design activities with the appeal of social innovation and the way of life and production advocated can successfully unify individual benefits with social and environmental benefits, so it is one of the ways to promote sustainable development. In the category of jewellery, returning to the sense of social responsibility, solving social problems and seeking social influence as the beginning of design can be the starting point of social innovation of jewellery, so as to propose more jewellery design solutions and achieve the balance between commercial value and social value.

Today, we jointly advocate using social responsibility to drive jewellery design innovation, integrate global design resources and collective wisdom, maintain an open attitude towards exchange and cooperation, accumulate experience, share resources, and jointly promote the bright development of jewellery design education!

CONTENTS

目录

展览现场

主展区
GALLERIES

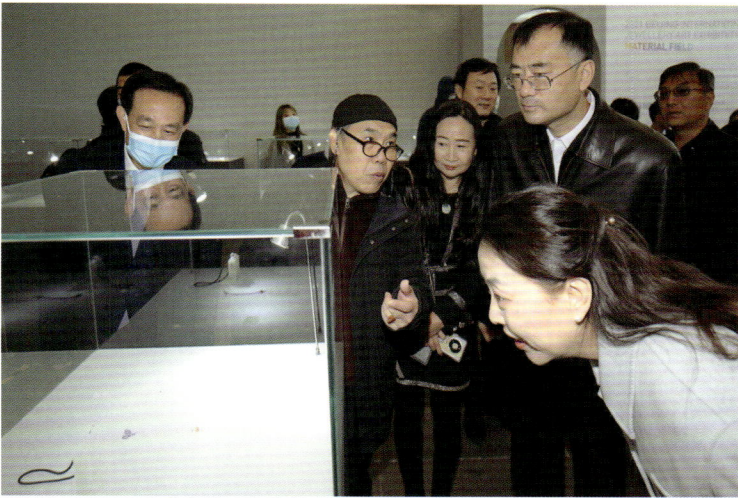

动态秀
RUNWAY SHOW

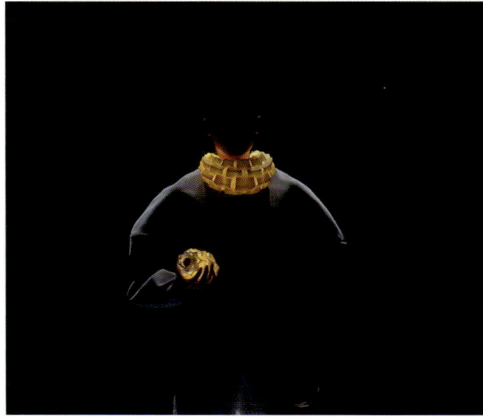

高峰论坛
ACADEMIC FORUM

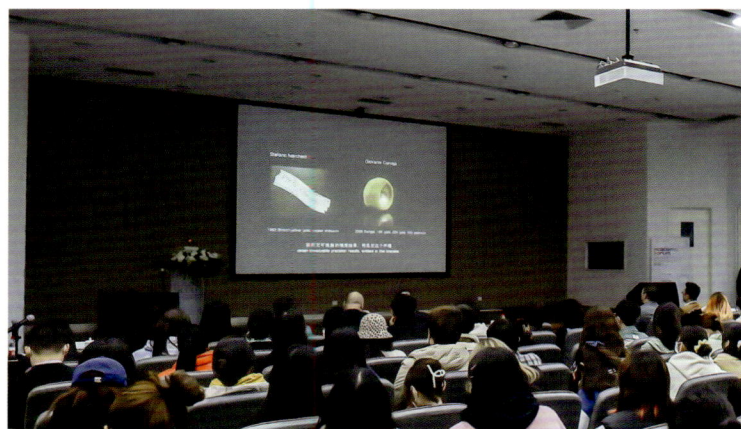

LIST OF ARTISTS

参展艺术家

2021 北京国际首饰艺术展

外国参展艺术家作品
WORKS OF OVERSEA ARTISTS

中国参展艺术家作品
WORKS OF CHINESE ARTISTS

一切成空
OUT OF NOTHING

作者姓名: Lital Mendel（以色列）
作品类型: 项链，胸针
作品材质: 银，锆石，环氧树脂

Artist: Lital Mendel（Israel）
Type: necklace , brooch
Material: silver, zirconiz, epoxy

砺行 / Honing

2021 北京国际
首饰艺术展
2021 BEIJING
INTERNATIONAL
JEWELLERY ART
EXHIBITION

67

前方施工，注意
ACHTUNG BAUSTELLE

作者姓名:　Maja Stojkovska（北马其顿共和国）
作品类型:　胸针，耳环
作品材质:　回收电脑散热器 —— 铝，回收电脑电缆，
　　　　　不锈钢，回收手机部件

Artist:　Maja Stojkovska
　　　　(The Republic of North Macedonia)
Type:　brooch, earring
Material:　reused computer hear sink — aluminium,
　　　　　reused computer cable, stainless steel,
　　　　　reused mobile phone parts

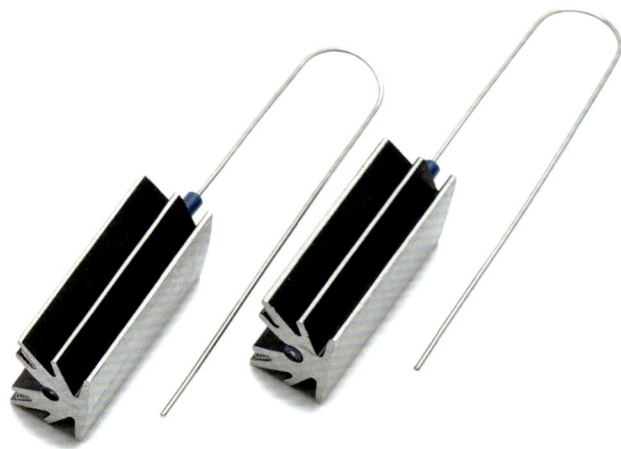

砺行 / Honing

2021 北京国际
首饰艺术展
2021 BEIJING
INTERNATIONAL
JEWELLERY ART
EXHIBITION

68

它越发靠近，但无人阻止
WHEN IT GOT CLOSER, NO ONE STOPPED

作者姓名：　Malene Kastalje（丹麦）
作品类型：　胸针
作品材质：　硅胶，钕磁铁，颜料

Artist:　　Malene Kastalje（Denmark）
Type:　　　brooch
Material:　 silicone, neodymium magnet, pigment

砺行 / Honing

2021 北京国际
首饰艺术展
2021 BEIJING
INTERNATIONAL
JEWELLERY ART
EXHIBITION

69

黑色时间 2，黑色六边形
BLACK TIME 2 , BLACK HEXAGON

作者姓名：　Maria Rosa Franzin（意大利）
作品类型：　胸针
作品材质：　银，金，甲基丙烯酸酯

Artist:　　Maria Rosa Franzin（Italy）
Type:　　　brooch
Material:　silver, gold, methacrylate

砺行 / Honing

2021 北京国际
首饰艺术展
2021 BEIJING
INTERNATIONAL
JEWELLERY ART
EXHIBITION

70

"变形"项链，"变形"手链，"变形"胸针
"METAMORPHOSIS" NECKLACE,
"METAMORPHOSIS" BRACELET,
"METAMORPHOSIS" BROOCH

作者姓名： Maria Tsimpiskaki（希腊）
作品类型： 项链，手镯，胸针
作品材质： 银，丝茧

Artist: Maria Tsimpiskaki（Greece）
Type: necklace, bracelet, brooch
Material: silver, silk cocoons

欲望森林；成长的心
FOREST OF DESIRE；GROWING HEART

作者姓名： Marina Zachou（希腊）
作品类型： 手镯
作品材质： 铜，金粉，尼龙材质印刷照片，清漆

Artist: Marina Zachou（Greece）
Type: bracelet
Material: copper, gold dust,
nylon printed photographs, varnish

砥行 / Honing

2021 北京国际
首饰艺术展
2021 BEIJING
INTERNATIONAL
JEWELLERY ART
EXHIBITION

71

剧院，蒙面，萌生
THEATRE, MASKED, INITIATION

作者姓名:　Marlene de Beer（南非）
作品类型:　耳环，项链
作品材质:　氧化 925 银，玻璃珠，棉线，陶瓷，925 银，珍珠，棉花

Artist:　　　Marlene de Beer（South Africa）
Type:　　　earring, necklace
Material:　　oxidised 925 silver, glass bead,
　　　　　　cotton string, porcelain, 925 silver, pearl, cotton

砺行 / Honing

2021 北京国际
首饰艺术展
2021 BEIJING
INTERNATIONAL
JEWELLERY ART
EXHIBITION

72

生命的迹象
SEGNI DI VITA (SIGN OF LIFE)

作者姓名: **Marta Alice Adda（意大利）**
作品类型: **胸针**
作品材质: **银**

Artist: **Marta Alice Adda (Italy)**
Type: **brooch**
Material: **silver**

砺行 / Honing

2021 北京国际
首饰艺术展
2021 BEIJING
INTERNATIONAL
JEWELLERY ART
EXHIBITION

73

生命的迹象
SEGNI DI VITA (SIGN OF LIFE)

作者姓名: Marta Alice Adda（意大利）
作品类型: 胸针
作品材质: 银

幼苗 I，幼苗 II，幼苗 III
SEEDLING I, SEEDLING II, SEEDLING III

作者姓名：　Marta Armada Rodríguez（西班牙）
作品类型：　项链
作品材质：　陶瓷，黄金，铂金，丝线
Artist:　　Marta Armada Rodríguez（Spain）
Type:　　　necklace
Material:　porcelain, gold, platinum, silk thread

砥行 / Honing

2021 北京国际
首饰艺术展
2021 BEIJING
INTERNATIONAL
JEWELLERY ART
EXHIBITION

74

海藻，流动，跳舞的珍珠
ALGAE, FLOWING, DANCING PEARLS

作者姓名： Marta Fernandez Caballero（西班牙）
作品类型： 戒指，项链
作品材质： 抛光青铜，氧化银，银，养殖珍珠

Artist: Marta Fernandez Caballero（Spain）
Type: ring, necklace
Material: polished bronze, patinated silver,
            silver, cultured pearl

砺行 / Honing

2021 北京国际
首饰艺术展
2021 BEIJING
INTERNATIONAL
JEWELLERY ART
EXHIBITION

75

图像之门，平静的生活 —— 蓝色回忆，有窗的房间
DOOR TO IMAGES, SILL LIFE—BLUE MEMORIES ,
ROOM WITH THE WINDOW

作者姓名: Namkyung Lee（韩国）
作品类型: 胸针
作品材质: 银，照片，印刷照片的亚克力，钻坯

Artist: Namkyung Lee（South Korea）
Type: brooch
Material: silver, photograph,
printed on acrylic, rough diamond

砺行 / Honing

2021 北京国际
首饰艺术展
2021 BEIJING
INTERNATIONAL
JEWELLERY ART
EXHIBITION

76

生活
ON LIFE

作者姓名:　Nicole Schuster（德国）
作品类型:　项链
作品材质:　氧化 925 银，18K 黄金，钛

Artist:　Nicole Schuster（Germany）
Type:　necklace
Material:　sterling silver oxidised,
　　　　　18K yellow gold, titanium

砺行 / Honing

2021 北京国际
首饰艺术展
2021 BEIJING
INTERNATIONAL
JEWELLERY ART
EXHIBITION

77

Kateřina Olivová 去看歌剧，项链 "我们的容器"
KATEŘINA OLIVOVÁ GOES TO THE OPERA,
NECKLACE "OUR CONTAINER"

作者姓名: Nicole Taubinger（捷克）
作品类型: 胸针
作品材质: 塑料垃圾，照片

Artist: Nicole Taubinger（The Czech Republic）
Type: brooch
Material: plastic waste, photograph

砺行 / Honing

2021 北京国际
首饰艺术展
2021 BEIJING
INTERNATIONAL
JEWELLERY ART
EXHIBITION

78

森林砍伐，
失落的迷宫花园与庇护所，
空气、水、土和火
DEFORESTATION,
LOST LABYRINTH GARDEN WITH SHELTER,
AIR, WATER, EARTH AND FIRE

作者姓名: Paolo Gambarelli（意大利）

作品类型: 戒指，胸针

作品材质: 镀钌黄铜／青铜及其氧化物，黄铜，磁铁，
带有矿物微聚集体的聚合树脂

Artist: Paolo Gambarelli（Italy）

Type: ring, brooch

Material: ruthenium plated brass/bronze, brass, magnets,
polymer resin with mineral micro-aggregates

2021 北京国际
首饰艺术展
2021 BEIJING
INTERNATIONAL
JEWELLERY ART
EXHIBITION

79

Ca-ta-ca Ⅰ（雅甘语，意思是 "来"）—— 2020，
Ca-ta-ca Ⅱ（雅甘语，意思是 "来"）—— 2020，
Ca-ta-ca Ⅲ（雅甘语，意思是 "来"）—— 2020

CA-TA-CA Ⅰ (YAGÁN TERM THAT MEANS COME) — 2020,
CA-TA-CA Ⅱ (YAGÁN TERM THAT MEANS COME) — 2020,
CA-TA-CA Ⅲ (YAGÁN TERM THAT MEANS COME) — 2020

作者姓名： Patricia Lglesias（智利）
作品类型： 项链，手镯，胸针
作品材质： 铜丝，剑麻，棉，麻，黄麻，棉丝，粘胶

Artist: Patricia Lglesias（Chile）
Type: necklace, Bracelet, Brooch
Material: copper wire, sisal, cotton, linen, jute,
cotton with silk, viscose

砺行 / Honing

2021 北京国际
首饰艺术展
2021 BEIJING
INTERNATIONAL
JEWELLERY ART
EXHIBITION

80

每日谷物，日常工具
DAILY GRAINS, DAILY TOOLS

作者姓名: Philipp Spillmann（瑞士）
作品类型: 项链
作品材质: 白色塑料筷子，人造红丝绳，
红色塑料筷子，塑料碗，丝线

Artist: Philipp Spillmann（Switzerland）
Type: necklace
Material: white plastic chopsticks, artificial red silk cord,
red plastic chopstick, plastic rice bowl, thread

砺行 / Honing

2021 北京国际
首饰艺术展
2021 BEIJING
INTERNATIONAL
JEWELLERY ART
EXHIBITION

81

作者姓名: Policarpova Nadejda（罗马尼亚）
作品类型: 胸针
作品材质: 涤纶和人造丝线，尼龙网，亚克力，不锈钢针

Artist: Policarpova Nadejda（Romania）
Type: brooch
Material: polyester and rayon threads, nylon mesh,
acrylics, stainless steel pin

砺行 / Honing

2021 北京国际
首饰艺术展
2021 BEIJING
INTERNATIONAL
JEWELLERY ART
EXHIBITION

82

银杏胸针"特里萨"，虞美人胸针"伊迪丝"，
玫瑰波洛领结"贾纳基"
GINKGO BROOCH "THERESE",
CORN POPPY BROOCH "EDITH",
ROSE BOLO TIE "JANAKI"

作者姓名：	Quintessenz B Reinhard Hampel & Veronique Cartier（法国）
作品类型：	胸针，其他
作品材质：	带锤纹的铜，火珐琅，3D 打印胸针底座，钢，黄铜，3D 打印波洛锁，皮革，石头
Artist:	Quintessenz B Reinhard Hampel & Veronique Cartier（France)
Type:	brooch, other
Material:	hammered copper, fire enamel, 3d printed brooch base, steel, brass, 3D printed bolo lock, leather, stones

关节，爆炸，圣塞巴斯蒂亚诺
SNODI, ESPLOSIONE, SAN SEBASTIANO

作者姓名：	Roberto Zanon（意大利）
作品类型：	手镯，戒指
作品材质：	EVA（乙烯醋酸乙烯酯）
Artist:	Roberto Zanon（Italy）
Type:	bracelet, ring
Material:	EVA (ethylene vinyl acetate)

砺行 / Honing

2021 北京国际
首饰艺术展
2021 BEIJING
INTERNATIONAL
JEWELLERY ART
EXHIBITION

83

心碎胸针，7 个小矮人心碎项链，泪珠
BROKEN HEART BROOCH, 7 DWARFS,
BREKEN HEART NECKLACE, TEARDROPS

作者姓名：　Ruta Naujalyte（立陶宛）

作品类型：　胸针，项链，耳环

作品材质：　织物，缝纫线

Artist:　　Ruta Naujalyte（Lithuania）

Type:　　brooch, necklace, earring

Material:　textile, sewing threads

砺行 / Honing

2021 北京国际
首饰艺术展

2021 BEIJING
INTERNATIONAL
JEWELLERY ART
EXHIBITION

84

放大，无题，短叶松
ZOOM IN, UNTITLED, JACK PINES

作者姓名：　Saerom Kong（韩国）
作品类型：　项链，胸针
作品材质：　水稻，氧化银，放大镜，树脂，白木

Artist:　　Saerom Kong（South Korea）
Type:　　 necklace, brooch
Material:　 rice, oxidized silver, magnify glass,
　　　　　 resin, white wood

砺行 / Honing

2021 北京国际
首饰艺术展
2021 BEIJING
INTERNATIONAL
JEWELLERY ART
EXHIBITION

85

蛋，七，日
EGG, SEVEN, SUN

作者姓名:　Sara Barbanti（意大利）
作品类型:　戒指，项链，胸针
作品材质:　木炭，银，金

Artist:　　Sara Barbanti（Italy）
Type:　　　ring, necklace, brooch
Material:　charcoal, silver, gold

砺行 / Honing

2021 北京国际
首饰艺术展
2021 BEIJING
INTERNATIONAL
JEWELLERY ART
EXHIBITION

86

干花，盛开的紫色
DRIED FLORAL, BLOOMING IN PURPLE

作者姓名:　Sara Shahak（以色列）

作品类型:　胸针

作品材质:　铁，黄铜，珐琅色，不锈钢，木炭，银

Artist:　Sara Shahak（Israel）

Type:　brooch

Material:　iron, brass, enamel colors,
stainless steel, charcoal, silver

砺行 / Honing

2021 北京国际
首饰艺术展
2021 BEIJING
INTERNATIONAL
JEWELLERY ART
EXHIBITION

87

崛起
THE RISE

作者姓名：　Seyma Esma Uygun（土耳其）
作品类型：　项链
作品材质：　塑料

Artist:　　Seyma Esma Uygun（Turkey）
Type:　　　necklace
Material:　plastic

关系
RELATIONS

作者姓名：　Sognando Lo Scirocco Jewels（意大利）
作品类型：　项链
作品材质：　棉花，苏格兰线，玻璃珠，半宝石和 X 光片

Artist:　　Sognando Lo Scirocco Jewels（Italy）
Type:　　　necklace
Material:　cotton, scotland thread, glass beads,
　　　　　　semi-precious stones and X-ray plates

砺行 / Honing

2021 北京国际
首饰艺术展
2021 BEIJING
INTERNATIONAL
JEWELLERY ART
EXHIBITION

88

我们 —— 我 1，我们 —— 我 2
WE — ME 1, WE — ME 2

作者姓名: Soonin Han（韩国）
作品类型: 项链
作品材质: 925 银，玻璃，镀金，陶瓷

Artist: Soonin Han（South Korea）
Type: necklace
Material: 925 silver, glass,
gold plating, porcelain

砺行 / Honing

2021 北京国际
首饰艺术展
2021 BEIJING
INTERNATIONAL
JEWELLERY ART
EXHIBITION

89

CWAB: LIEN #22
CWAB: LIEN #22

作者姓名: Teresa F Faris（美国）
作品类型: 项链
作品材质: 纯银，桦木，鸟类雕琢的木材，
回收棉绳（Comfy Perch ™）

Artist: Teresa F Faris（America）
Type: necklace
Material: sterling silver, birch,
wood altered by a bird,
reclaimed cotton rope (Comfy Perch ™)

砺行 / Honing

2021 北京国际
首饰艺术展
2021 BEIJING
INTERNATIONAL
JEWELLERY ART
EXHIBITION

90

障碍
BARRIERS

作者姓名: Valentijn Vanmeirhaeghe（比利时）
作品类型: 其他
作品材质: 铁，纯银

Artist: Valentijn Vanmeirhaeghe (Belgium)
Type: other
Material: iron, sterling silver

保管（1）
SAFEKEEPING (1)

作者姓名: Veronica Cheann（挪威）
作品类型: 项链
作品材质: 烧烤木炭，棉线，
用水仙花（植物）染色的棉线

Artist: Veronica Cheann（Norway）
Type: necklace
Material: grill charcoal, cotton thread,
cotton thread colored with narcisser (plant)

砥行 / Honing

2021北京国际
首饰艺术展
2021 BEIJING
INTERNATIONAL
JEWELLERY ART
EXHIBITION

91

分子循环，暮光小交响曲
MOLECULAR CYCLE, TWILIGHT SINFONIA

作者姓名： Viktoria Münzker（斯洛伐克）

作品类型： 胸针，项链

作品材质： 浮木，阳极氧化钛，珍珠，玻璃，纯银，清漆，
黄铜，蓝宝石，水晶，锥体，微粒

Artist: Viktoria Münzker（Slovakia）

Type: brooch, necklace

Material: driftwood, anodized titanium, pearl, glass,
sterling silver, lacquer, brass, sapphires,
rock crystal, cone, micro-granules

砺行 / Honing

2021 北京国际
首饰艺术展

2021 BEIJING
INTERNATIONAL
JEWELLERY ART
EXHIBITION

92

关爱地球 #1，关爱地球 #2，关爱地球 #4
EARTH CARE #1, EARTH CARE #2, EARTH CARE #4

作者姓名: Yasmin Vinograd（以色列）
作品类型: 其他
作品材质: 银，铁，铜，铜绿，不锈钢，合成树脂，油漆

Artist: Yasmin Vinograd（Israel）
Type: other
Material: silver, iron, copper, patina, stainless steel, synthetic resin, paint

砥行 / Honing

2021 北京国际
首饰艺术展
2021 BEIJING
INTERNATIONAL
JEWELLERY ART
EXHIBITION

93

圣甲虫
SCARABAEUS SACER

作者姓名: Ylenia Deriu（意大利）
作品类型: 项链
作品材质: 环氧树脂，煤，浮石，颜料，氧化铜，
水性清漆，尼龙线，磁铁

Artist: Ylenia Deriu（Italy）
Type: necklace
Material: epoxy resin, coal, pumice stone, pigments,
oxidised brass, waterbased clear laquer,
nylon line, magnet

砺行 / Honing

2021 北京国际
首饰艺术展
2021 BEIJING
INTERNATIONAL
JEWELLERY ART
EXHIBITION

94

WORKS OF CHINESE ARTISTS

中国参展艺术家作品

时空 ——虫洞
TIME AND SPACE — WORMHOLES

作者姓名: 蔡蔚然（中国）
作品类型: 项链
作品材质: 铜

Artist: Cai Weiran（China）
Type: necklace
Material: copper

稚气
CHILDISHNESS

作者姓名: 蔡紫璞（中国）
作品类型: 胸针
作品材质: 925 银，冷珐琅，镜面纸，云马线，天鹅绒

Artist: Cai Zipu（China）
Type: brooch
Material: 925 silver, cold enamel, mirror paper, cloud horse wire, velvet

砺行 / Honing

2021 北京国际
首饰艺术展
2021 BEIJING
INTERNATIONAL
JEWELLERY ART
EXHIBITION

98

层叠·皱法
LAMINATED · WRINKLE METHOD

作者姓名: 曾博（中国）
作品类型: 项链
作品材质: 陶瓷，黄铜，棉绳

Artist: Zeng Bo（China）
Type: necklace
Material: porcelain, brass, cotton cord

浮华乱世
A WORLD OF POMP AND CHAOS

作者姓名: 曾嫚，赵猛（中国）
作品类型: 胸针
作品材质: 925 银，铜，大漆，螺钿

Artist: Zeng Man, Zhao Meng（China）
Type: brooch
Material: 925 silver, copper, lacquer, inlay

砺行 / Honing

2021 北京国际
首饰艺术展
2021 BEIJING
INTERNATIONAL
JEWELLERY ART
EXHIBITION

99

作者姓名： 曾义平（中国）

作品类型： 胸针

作品材质： 乌木，马贝珍珠，银

Artist:　　Zeng Yiping（China）

Type:　　　brooch

Material:　ebony, mabe pearl, silver

砺行 / Honing

2021 北京国际
首饰艺术展

2021 BEIJING
INTERNATIONAL
JEWELLERY ART
EXHIBITION

100

丛
SERIES

作者姓名:　陈宝瑜（中国）
作品类型:　胸针
作品材质:　银，红铜

Artist:　　**Chen Baoyu（China）**
Type:　　　**brooch**
Material:　**silver, red copper**

砥行 / Honing

2021 北京国际
首饰艺术展
2021 BEIJING
INTERNATIONAL
JEWELLERY ART
EXHIBITION

101

想要坐在你身边
WANTING TO SIT NEXT TO YOU

作者姓名: 陈达礼（中国）
作品类型: 胸针
作品材质: 银镀金

Artist: Chen Dali（China）
Type: brooch
Material: gold plated silver

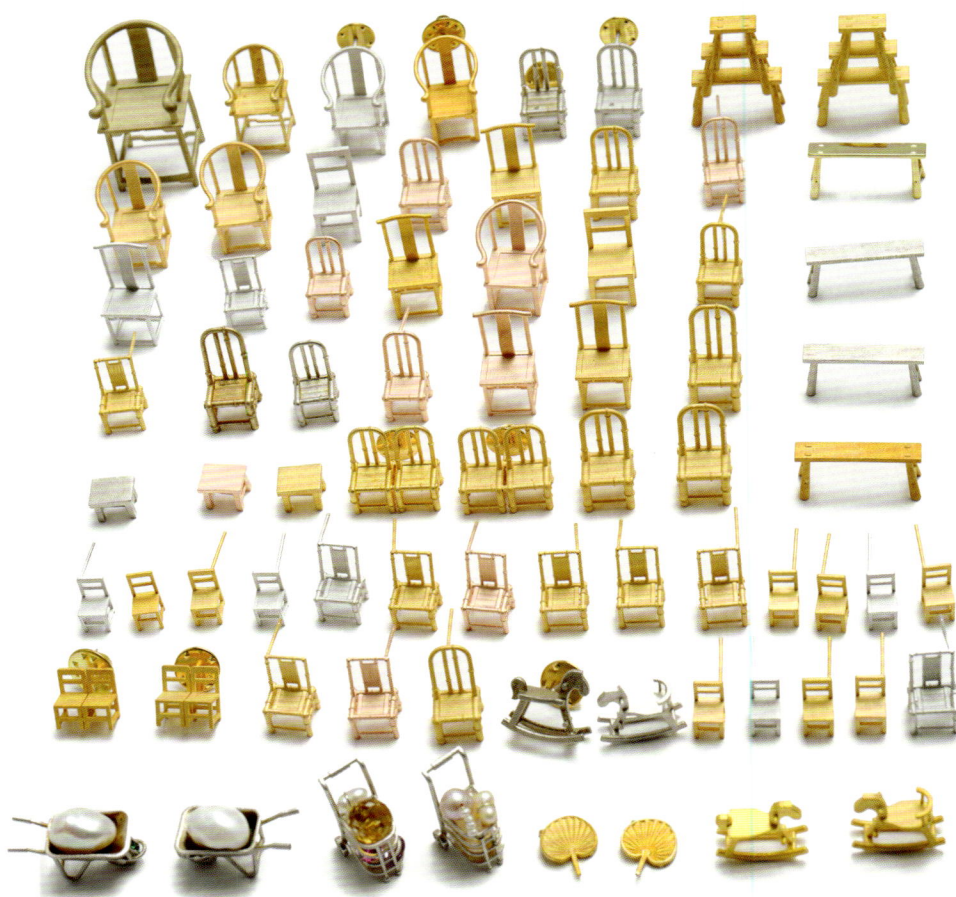

砺行 / Honing

2021 北京国际
首饰艺术展
2021 BEIJING
INTERNATIONAL
JEWELLERY ART
EXHIBITION

102

心灵感应
TELEPATHIC

作者姓名:　陈婧娴（中国）
作品类型:　戒指
作品材质:　硅胶管，银

Artist:　　Chen Jingxian（China）
Type:　　　ring
Material:　 silicone tube, silver

砺行 / Honing

2021 北京国际
首饰艺术展
2021 BEIJING
INTERNATIONAL
JEWELLERY ART
EXHIBITION

103

蜕变
TRANSFORMATION

作者姓名: 　陈静（中国）
作品类型: 　其他
作品材质: 　绿松石边角料，925 银

Artist: 　　Chen Jing（China）
Type: 　　　other
Material: 　turquoise leftovers, sterling silver

侗·饰
DONG-TRIM

作者姓名: 　陈琳琳（中国）
作品类型: 　胸针
作品材质: 　侗布，925 银，不锈钢线

Artist: 　　Chen Linlin（China）
Type: 　　　brooch
Material: 　Dong cloth, 925 silver, stainless steel wire

砥行 / Honing

2021 北京国际
首饰艺术展

2021 BEIJING
INTERNATIONAL
JEWELLERY ART
EXHIBITION

104

熵增
ENTROPY INCREASE

作者姓名: 陈敏（中国）
作品类型: 项链
作品材质: 锡，珐琅

Artist: Chen Min（China）
Type: necklace
Material: tin, enamel

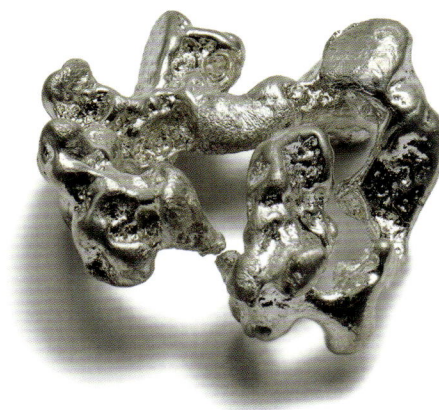

月光之诗系列 —— 银丝蕾丝编织手链
MOONLIGHT POETRY COLLECTION — SILVER LACE BRAID BRACELET

作者姓名: 陈尚仪（中国）
作品类型: 手镯
作品材质: 999 银，珍珠

Artist: Chen Shangyi（China）
Type: bracelet
Material: 999 silver, pearl

砺行 / Honing

2021 北京国际
首饰艺术展
2021 BEIJING
INTERNATIONAL
JEWELLERY ART
EXHIBITION

105

三生万物
THE THREE CREATURES

作者姓名： 陈怡（中国）
作品类型： 胸针
作品材质： 银镀金

Artist: Chen Yi（China）
Type: brooch
Material: gold plated silver

睡梦系列 2
SLEEPING SERIES 2

作者姓名： 陈艺，闯响（中国）
作品类型： 胸针
作品材质： 925 银，植鞣皮，黄铜

Artist: Chen Yi, Chuang Xiang（China）
Type: brooch
Material: 925 silver, vegetable tanned leather, brass

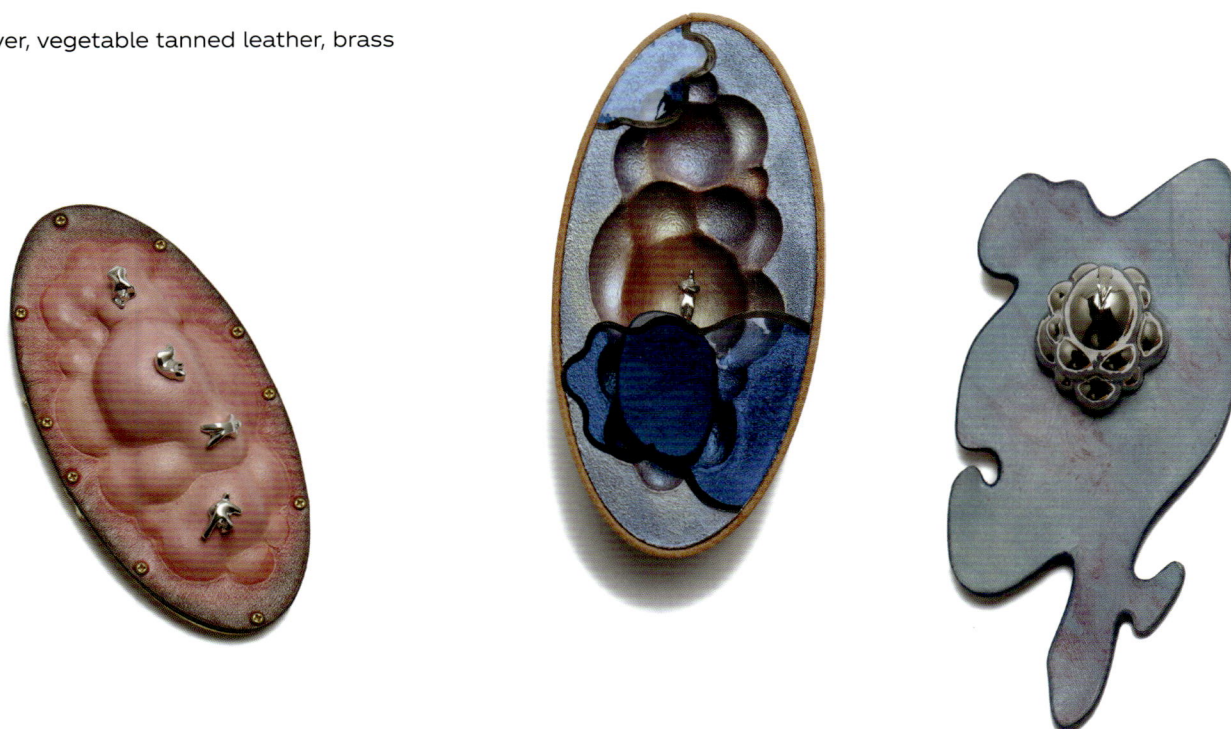

砺行 / Honing

2021 北京国际
首饰艺术展
2021 BEIJING
INTERNATIONAL
JEWELLERY ART
EXHIBITION

106

十四
FOURTEEN

作者姓名: 陈玉欣（中国）
作品类型: 胸针
作品材质: 黑木，紫铜，银，珐琅

Artist: Chen Yuxin（China）
Type: brooch
Material: blackwood, bronze, silver, enamel

洞庭珍珠
DONGTING PEARLS

作者姓名: 陈昱（中国）
作品类型: 项链
作品材质: 洞庭湖产糯米，银

Artist: Chen Yu （China）
Type: necklace
Material: Dongting Lake produces glutinous rice, silver

砺行 / Honing

2021 北京国际
首饰艺术展
2021 BEIJING
INTERNATIONAL
JEWELLERY ART
EXHIBITION

107

厝檐 1——综合材料胸针
ALCOVE 1 — COMPOSITE MATERIAL BROOCH

作者姓名: **程滢（中国）**
作品类型: **胸针**
作品材质: **软木，大漆，珐琅，银**

Artist: **Cheng Ying（China）**
Type: **brooch**
Material: **cork, lacquer, enamel, silver**

砺行 / Honing

2021 北京国际
首饰艺术展
2021 BEIJING
INTERNATIONAL
JEWELLERY ART
EXHIBITION

108

大自在荷——夏雨荷，大自在荷——一品清莲，
大自在荷——鱼戏，大自在荷——月下独赏
ENJOY GREAT FREEDOM—NATURAL LOTUS,
LOTUS LEAF IN SUMMER RAIN,
A PINT OF FRESH LOTUS,
FISH PLAY AMONG THE LOTUS LEAVE,
YOU CAN HAVE IT ALL UNDER THE MOON

作者姓名： 程园（中国）
作品类型： 胸针，戒指
作品材质： 尖晶石，欧泊，钻石，18K 黄金，
　　　　　翡翠，和田玉

Artist:　　Cheng Yuan（China）
Type:　　　brooch, ring
Material:　spinel, opal, diamonds, 18K yellow gold,
　　　　　emerald, nephrite

砺行 / Honing

2021 北京国际
首饰艺术展
2021 BEIJING
INTERNATIONAL
JEWELLERY ART
EXHIBITION

109

晨祷
MORNING PRAYERS

作者姓名： 戴夏涵（中国）
作品类型： 项链
作品材质： 黄铜，银，紫光檀，树脂，水晶线

Artist: Dai Xiahan（China）
Type: necklace
Material: brass, silver, red sandalwood, resin, crystal thread

2021 北京国际
首饰艺术展

2021 BEIJING
INTERNATIONAL
JEWELLERY ART
EXHIBITION

作者姓名： 戴翔（中国）

作品类型： 胸针

作品材质： 木材，木皮，中国画岩彩，不锈钢针

Artist: Dai Xiang（China）

Type: brooch

Material: wood, wood veneer,
China painted rock colour, stainless steel pins

砺行 / Honing

2021 北京国际
首饰艺术展
2021 BEIJING
INTERNATIONAL
JEWELLERY ART
EXHIBITION

111

保护
PROTECTION

作者姓名: 邓婧怡（中国）
作品类型: 项链
作品材质: 银，橡胶，海洋玛瑙

Artist: Deng Jingyi（China）
Type: necklace
Material: silver, rubber, marine agate

镜子 2
MIRROR 2

作者姓名: 丁雪妍（中国）
作品类型: 胸针
作品材质: 925 银，化妆海绵，眼影

Artist: Ding Xueyan（China）
Type: brooch
Material: 925 silver, cosmetic puff, eye shadow

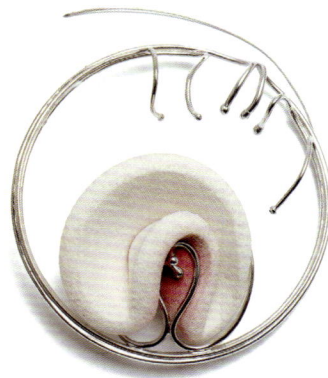

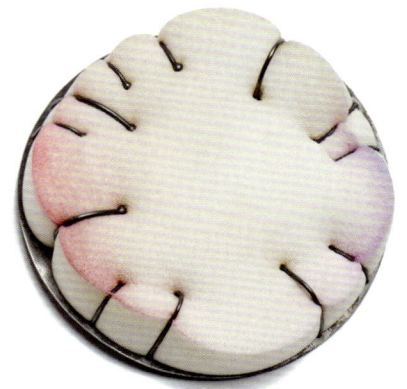

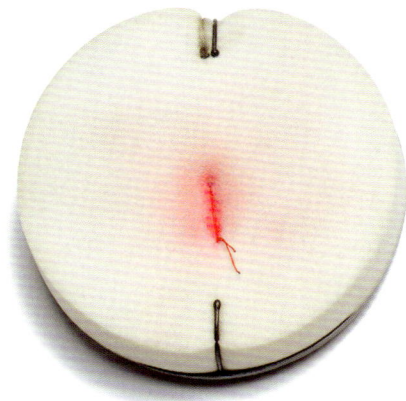

砺行 / Honing

2021 北京国际
首饰艺术展
2021 BEIJING
INTERNATIONAL
JEWELLERY ART
EXHIBITION

112

亭
PAVILION

作者姓名: 董靖阳（中国）
作品类型: 其他
作品材质: 925 银，合成立方氧化锆

Artist: Dong Jingyang（China）
Type: other
Material: 925 silver, synthetic cubic zirconia

亲密
CLOSENESS

作者姓名: 董敏媛（中国）
作品类型: 项链
作品材质: 黄铜丝，硅胶，沙子，树刺，木板

Artist: Dong Minyuan（China）
Type: necklace
Material: brass wire, silicone,
sand, tree thorns, planks

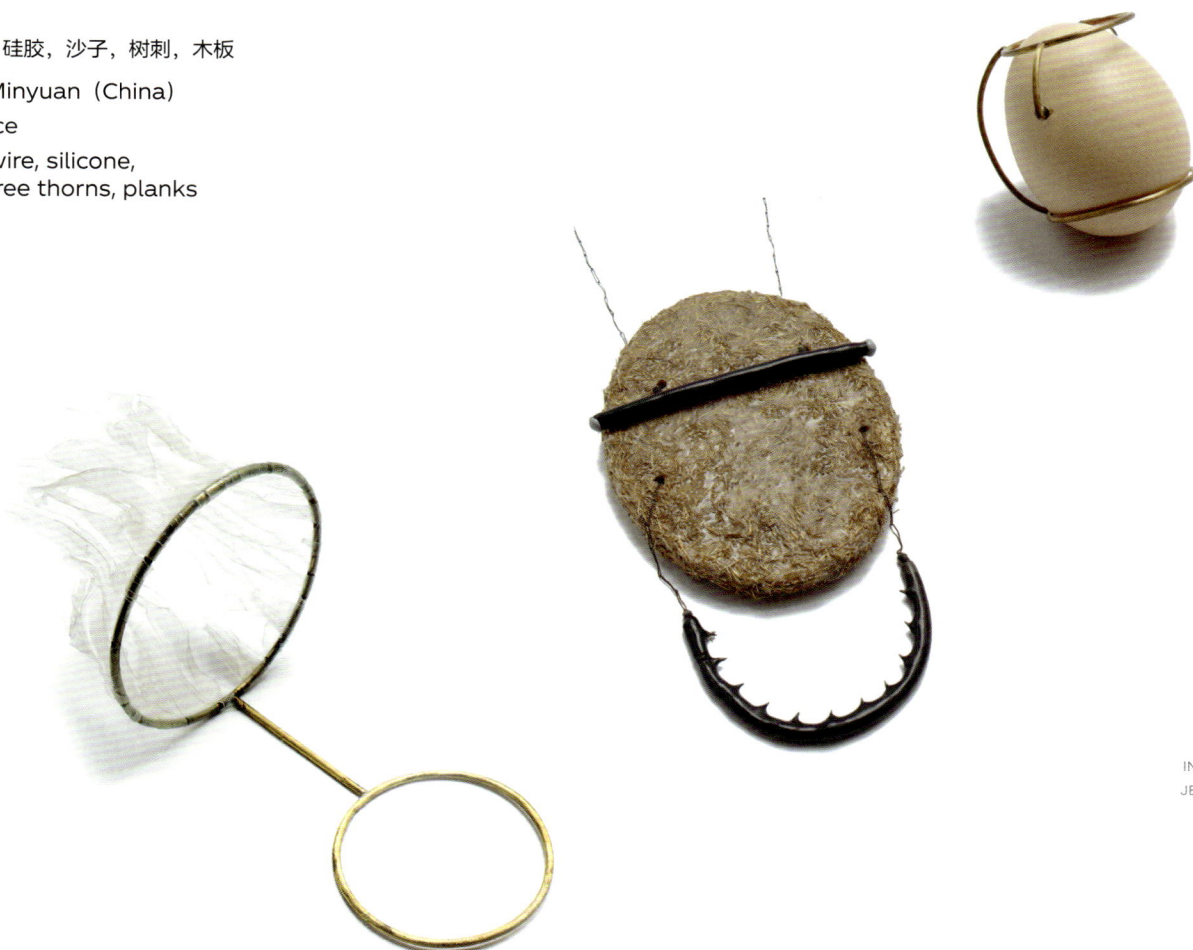

砺行 / Honing

2021 北京国际
首饰艺术展
2021 BEIJING
INTERNATIONAL
JEWELLERY ART
EXHIBITION

113

"墨·白"系列作品 1——手镯
INK·WHITE COLLECTION 1 — BRACELET

作者姓名:　豆肖楠（中国）
作品类型:　手镯
作品材质:　墨玉（带有天然黄铁矿伴生矿），珍珠，18K 黄金

Artist:　　Dou Xiaonan（China）
Type:　　　bracelet
Material:　dark jade (associated with pyrite),
　　　　　 pearls, 18K yellow gold

光域系列 2
LIGHTFIELD SERIES 2

作者姓名:　杜妍（中国）
作品类型:　胸针
作品材质:　珐琅银，箔，青铜，珍珠，亚克力

Artist:　　Du Yan（China）
Type:　　　brooch
Material:　silver enamel, foil,
　　　　　 bronze, pearl, acrylic

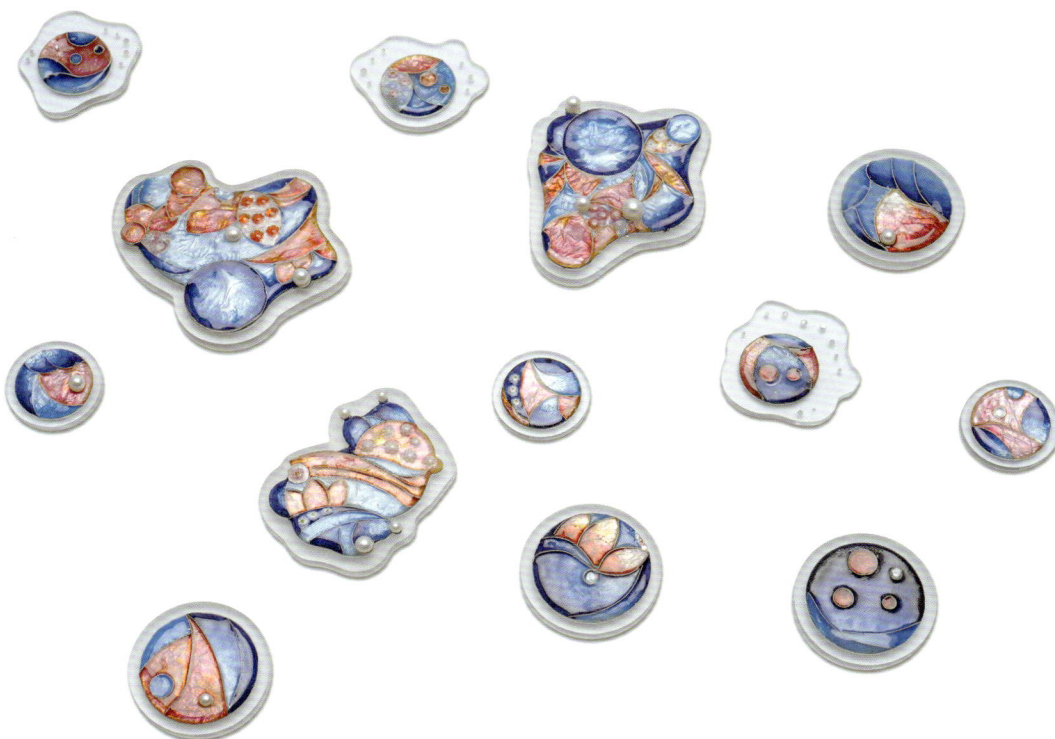

硕行 / Honing

2021 北京国际
首饰艺术展
2021 BEIJING
INTERNATIONAL
JEWELLERY ART
EXHIBITION

114

跃 2020
LEAP IN 2020

作者姓名: 段丙文（中国）
作品类型: 胸针
作品材质: 银

Artist: Duan Bingwen（China）
Type: brooch
Material: silver

砺行 / Honing

2021 北京国际
首饰艺术展
2021 BEIJING
INTERNATIONAL
JEWELLERY ART
EXHIBITION

115

作者姓名: 段燕俪（中国）
作品类型: 胸针
作品材质: 银，古瓷片

Artist: Duan Yanli（China）
Type: brooch
Material: silver, ancient ceramic chip

砺行 / Honing

2021 北京国际
首饰艺术展
2021 BEIJING
INTERNATIONAL
JEWELLERY ART
EXHIBITION

116

下一刻，奇迹！
THE NEXT MOMENT, A MIRACLE!

作者姓名： 范文琦（中国）
作品类型： 胸针
作品材质： 925 银，14K 黄金，冷珐琅，合成立方氧化锆

Artist: Fan Wenqi（China）
Type: brooch
Material: 925 silver, 14K gold, cold enamel,
synthetic cubic zirconia

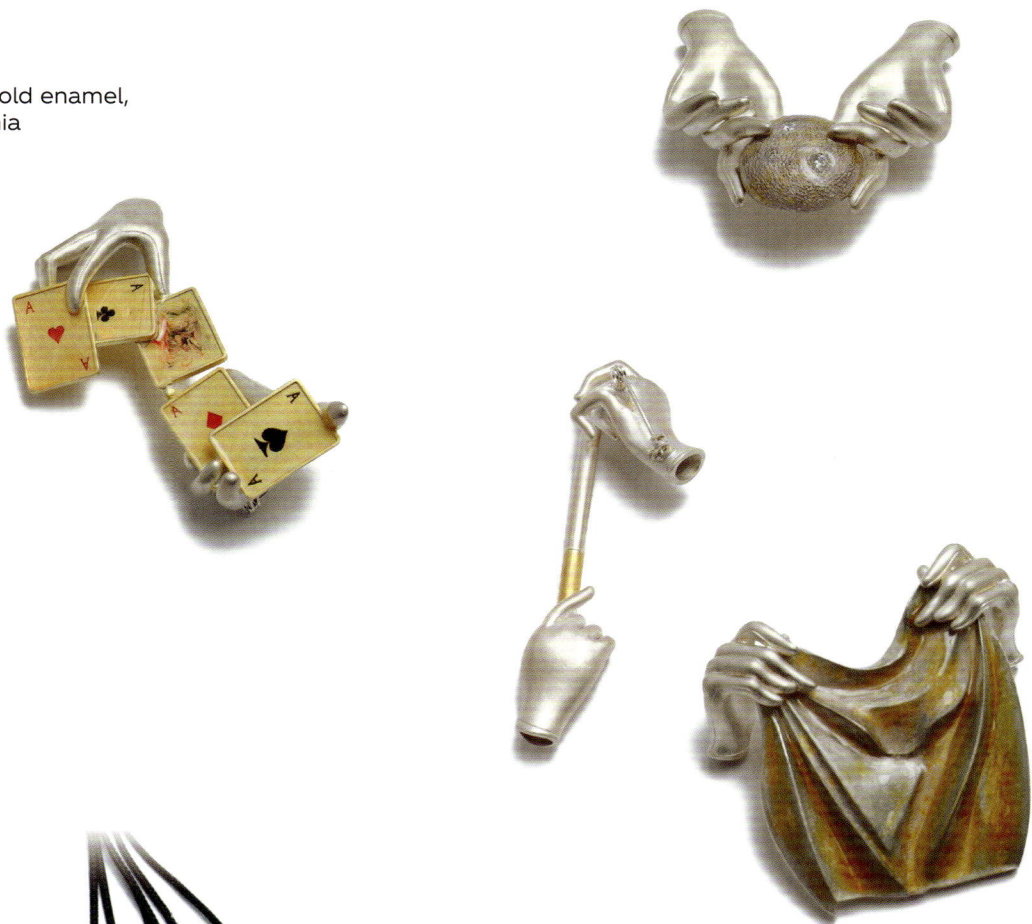

再一次 3
ONCE MORE 3

作者姓名： 方修（中国）
作品类型： 项链
作品材质： 银

Artist: Fang Xiu（China）
Type: necklace
Material: silver

砺行 / Honing

2021 北京国际
首饰艺术展
2021 BEIJING
INTERNATIONAL
JEWELLERY ART
EXHIBITION

117

双时之间
THE DOUBLE HOUR

作者姓名: 冯剑逸（中国）
作品类型: 项链
作品材质: 阳极氧化铝，银，钛，锆石

Artist:　　Feng Jianyi（China）
Type:　　　necklace
Material:　anodised aluminium, silver, titanium, zirconia

砺行 / Honing

2021 北京国际
首饰艺术展

2021 BEIJING
INTERNATIONAL
JEWELLERY ART
EXHIBITION

118

记·忆——清末
MEMORIES—LATE QING DYNASTY

作者姓名: 冯诗艺（中国）
作品类型: 胸针
作品材质: 银，银镀金，仿羊皮

Artist: Feng Shiyi（China）
Type: brooch
Material: silver, gold plated silver, imitation sheepskin

情绪消化器
EMOTIONAL DIGESTERS

作者姓名: 付家琪（中国）
作品类型: 项链
作品材质: 海绵，琉璃，树脂

Artist: Fu Jiaqi（China）
Type: necklace
Material: sponge, glaze, resin

砺行 / Honing

2021 北京国际
首饰艺术展
2021 BEIJING
INTERNATIONAL
JEWELLERY ART
EXHIBITION

119

姑且叫它山石
WE SHALL CALL IT A VALUABLE STONE

作者姓名: 傅永和（中国）
作品类型: 项链
作品材质: 铜，珐琅

Artist: Fu Yonghe（China）
Type: necklace
Material: copper, enamel

砺行 / Honing

2021 北京国际
首饰艺术展
2021 BEIJING
INTERNATIONAL
JEWELLERY ART
EXHIBITION

120

非凡的仪式——绣球
EXTRAORDINARY RITUALS — STRING BALL

作者姓名: 甘滔（中国）
作品类型: 其他
作品材质: 黄铜，布，麦秆，碧玉

Artist: Gan Tao（China）
Type: other
Material: brass, cloth, straw and jasper

雕金唐草纹手镯
GILDED TANG DYNASTY FLORAL DESIGN BRACELET

作者姓名: 高静（中国）
作品类型: 手镯
作品材质: 银

Artist: Gao Jing（China）
Type: bracelet
Material: silver

砺行 / Honing

2021 北京国际
首饰艺术展
2021 BEIJING
INTERNATIONAL
JEWELLERY ART
EXHIBITION

121

"异像" 系列
THE VISION SERIES

作者姓名: 高珊（中国）
作品类型: 项链
作品材质: 925 银，玻璃成品，金箔

Artist: Gao Shan（China）
Type: necklace
Material: 925 silver, finished glass, gold leaf

砺行 / Honing

2021 北京国际
首饰艺术展
2021 BEIJING
INTERNATIONAL
JEWELLERY ART
EXHIBITION

122

一生三——游
ONE LIFE THREE — TOUR

作者姓名: **高伟（中国）**
作品类型: **其他**
作品材质: **银**

Artist: **Gao Wei（China）**
Type: **other**
Material: **silver**

砺行 / Honing

2021 北京国际
首饰艺术展
2021 BEIJING
INTERNATIONAL
JEWELLERY ART
EXHIBITION

123

叽叽喳喳
CHIRP

作者姓名: 高艺司（中国）
作品类型: 胸针
作品材质: 墨翠，925 银

Artist: Gao Yisi（China）
Type: brooch
Material: black gade, 925 silver

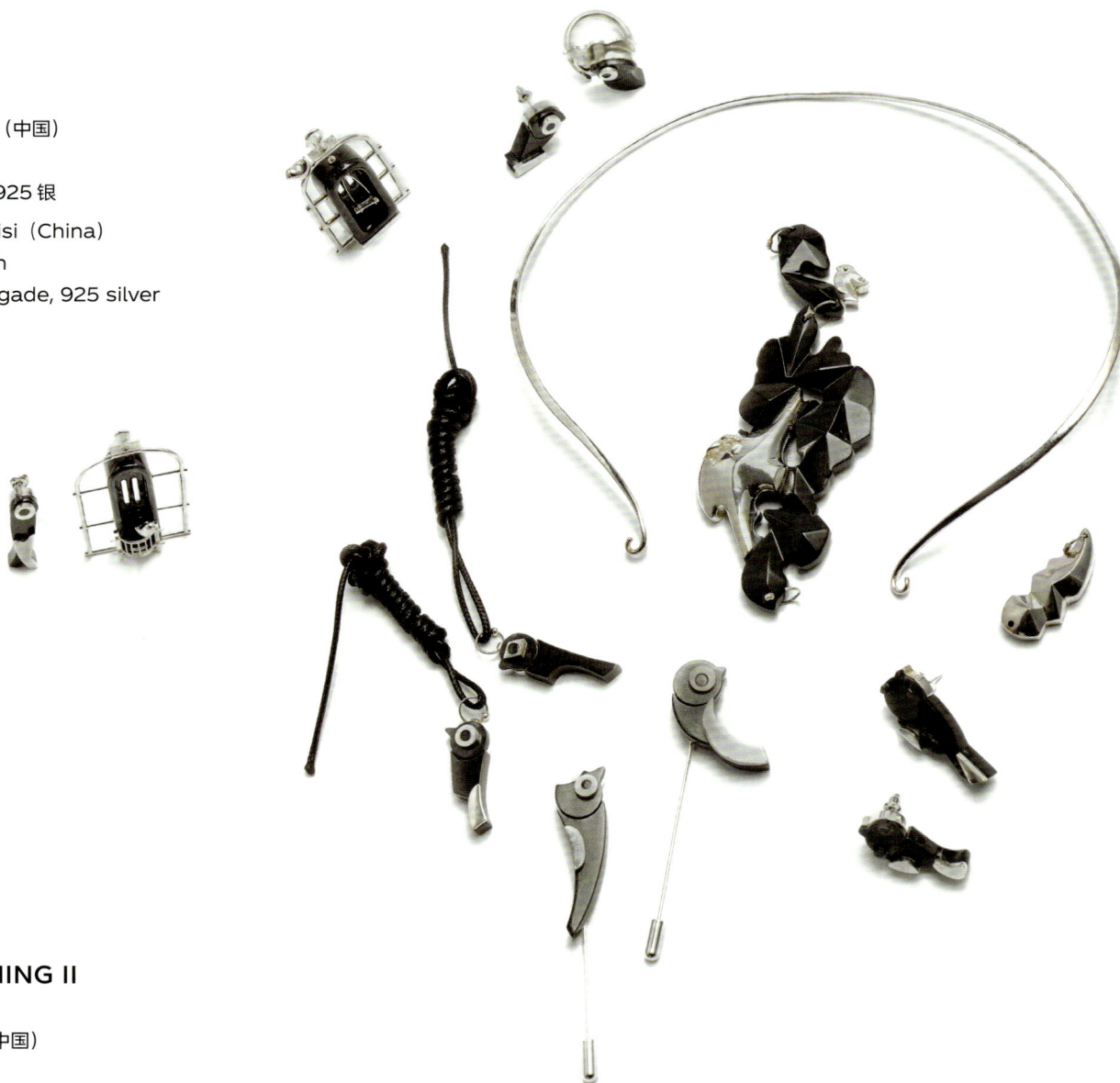

心动之二
HEARTWARMING II

作者姓名: 葛韵（中国）
作品类型: 胸针
作品材质: 黄铜，树脂，油漆，亚克力

Artist: Ge Yun（China）
Type: brooch
Material: brass, resin, paint, acrylic

砺行 / Honing

2021 北京国际
首饰艺术展
2021 BEIJING
INTERNATIONAL
JEWELLERY ART
EXHIBITION

124

月下 1, 月下 2
UNDER THE MOON 1, UNDER THE MOON 2

作者姓名:　宫婷（中国）
作品类型:　胸针
作品材质:　玉，18K 黄金，钻石

Artist:　　Gong Ting（China）
Type:　　　brooch
Material:　jade, 18K yellow gold, diamonds

2021 北京国际
首饰艺术展
2021 BEIJING
INTERNATIONAL
JEWELLERY ART
EXHIBITION

125

作者姓名:　宫婷（中国）
作品类型:　胸针

断鸿声里
HEARING A LONELY SWAN'S SONG

作者姓名：　**龚时雨（中国）**
作品类型：　**胸针**
作品材质：　**赤铜，银，大漆，檀木**

Artist:　　**Gong Shiyu（China）**
Type:　　　**brooch**
Material:　**copper, silver, lacquer, sandalwood**

砺行 / Honing

2021 北京国际
首饰艺术展
2021 BEIJING
INTERNATIONAL
JEWELLERY ART
EXHIBITION

126

石榴 · 花儿 · 女子，银窗花，水中石
POMEGRANATE — FLOWER — WOMAN,
SILVER GRILLE, STONE IN THE WATER

作者姓名：　古丽米拉 · 艾尼（中国）

作品类型：　胸针，项链

作品材质：　3D 打印尼龙喷漆，金属，珍珠，青金石，
　　　　　　925 银，和田玉

Artist:　　　Gulimila Aini（China）

Type:　　　brooch

Material:　　3D printed nylon spray paint, metal,
　　　　　　pearl, lapis lazuli, 925 silver, nephrite

砺行 / Honing

2021 北京国际
首饰艺术展
2021 BEIJING
INTERNATIONAL
JEWELLERY ART
EXHIBITION

127

以主之名
IN THE NAME OF THE LORD

作者姓名： 谷明（中国）
作品类型： 项链
作品材质： 钧瓷，银，金丝海柳，珍珠，石榴石

Artist: Gu Ming（China）
Type: necklace
Material: Jun porcelain, silver,
golden black coral, pearls, garnet

"盾" 系列 2
THE SHIELD SERIES 2

作者姓名： 顾浩（中国）
作品类型： 胸针
作品材质： 银，不锈钢，板栗壳，乌龟脊椎

Artist: Gu Hao（China）
Type: brooch
Material: silver, stainless steel,
chestnut shell, turtle spine

砥行 / Honing

2021 北京国际
首饰艺术展
2021 BEIJING
INTERNATIONAL
JEWELLERY ART
EXHIBITION

128

药言妙道
THE WONDERFUL WORD OF MEDICINE

作者姓名： 顾慧霞（中国）
作品类型： 胸针
作品材质： 925 银，玻璃，蜜蜡，中药材

Artist: Gu Huixia（China）
Type: brooch
Material: 925 silver, glass, beeswax and herbs

改变 · 控制
CHANGE — CONTROL

作者姓名： 顾悦（中国）
作品类型： 胸针
作品材质： 925 银，黄铜，紫铜，亚克力板

Artist: Gu Yue（China）
Type: brooch
Material: 925 silver, brass,
purple copper, acrylic plate

砺行 / Honing

2021 北京国际
首饰艺术展
2021 BEIJING
INTERNATIONAL
JEWELLERY ART
EXHIBITION

129

近乎标准 1，近乎标准 2
NEAR CRITERION 1, NEAR CRITERION 2

作者姓名： 关宇洋（中国）
作品类型： 胸针
作品材质： 尼龙，光敏树脂，925 银，锆石

Artist: Guan Yuyang（China）
Type: brooch
Material: nylon, photosensitive resin,
925 silver, zircon

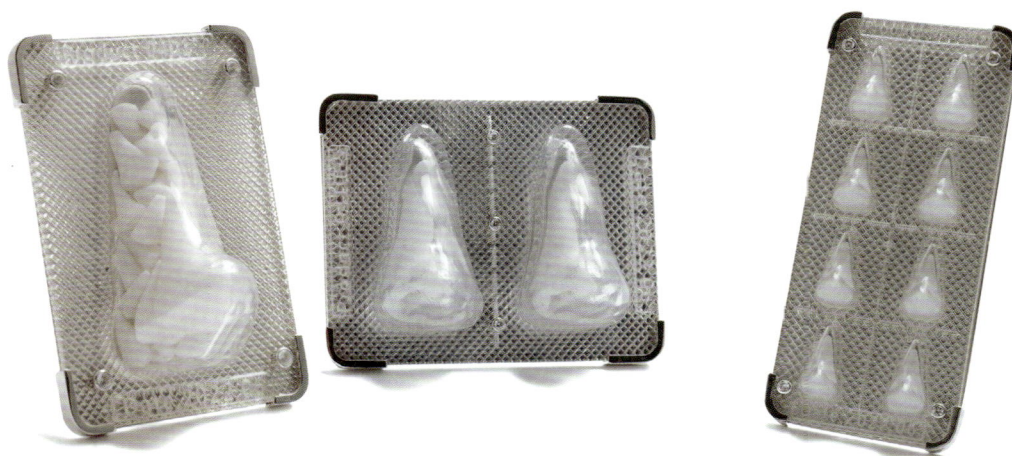

疫·痕
EPIDEMIC-TRACE

作者姓名： 郭琛宇（中国）
作品类型： 其他
作品材质： 925 银，大漆

Artist: Guo Chenyu（China）
Type: other
Material: 925 silver, lacquer

砺行 / Honing

2021 北京国际
首饰艺术展

2021 BEIJING
INTERNATIONAL
JEWELLERY ART
EXHIBITION

130

秀系列
SHOW

作者姓名： 郭鸿旭（中国）

作品类型： 耳饰，项链，手镯

作品材质： 银镀白金，刺绣，紫水晶，锆石

Artist: Guo Hongxü（China）

Type: earring, necklace, bracelet

Material: silver plated white gold, embroidered, amethyst, zirconia

新生
NEWBORN

作者姓名： 郭靖凯（中国）

作品类型： 戒指

作品材质： 18K 黄金，18K 白金，莫桑钻，蓝色月光石，钻石

Artist: Guo Jingkai（China）

Type: ring

Material: 18K yellow gold, 18K white gold, moissanite, blue moonstone, diamonds

砺行 / Honing

2021 北京国际
首饰艺术展
2021 BEIJING
INTERNATIONAL
JEWELLERY ART
EXHIBITION

131

紫荆花与鱼
CERCIS CHINENSIS AND FISH

作者姓名: 　郭莉莉（中国）
作品类型: 　胸针
作品材质: 　钛金属，18K 黄金，水晶，钻石

Artist: 　Guo Lili（China）
Type: 　brooch
Material: 　titanium, 18K yellow gold, crystal, diamonds

砺行 / Honing

2021 北京国际
首饰艺术展
2021 BEIJING
INTERNATIONAL
JEWELLERY ART
EXHIBITION

132

白描系列 #2
WHITE DRAWING SERIES #2

作者姓名: 郭新（中国）
作品类型: 胸针
作品材质: 银，画珐琅

Artist: Guo Xin（China）
Type: brooch
Material: silver, painted enamel

砺行 / Honing

2021 北京国际
首饰艺术展
2021 BEIJING
INTERNATIONAL
JEWELLERY ART
EXHIBITION

133

作者姓名: 韩琦（中国）
作品类型: 其他
作品材质: 银，玻璃，沙

Artist: Han Qi（China）
Type: other
Material: silver, glass, sand

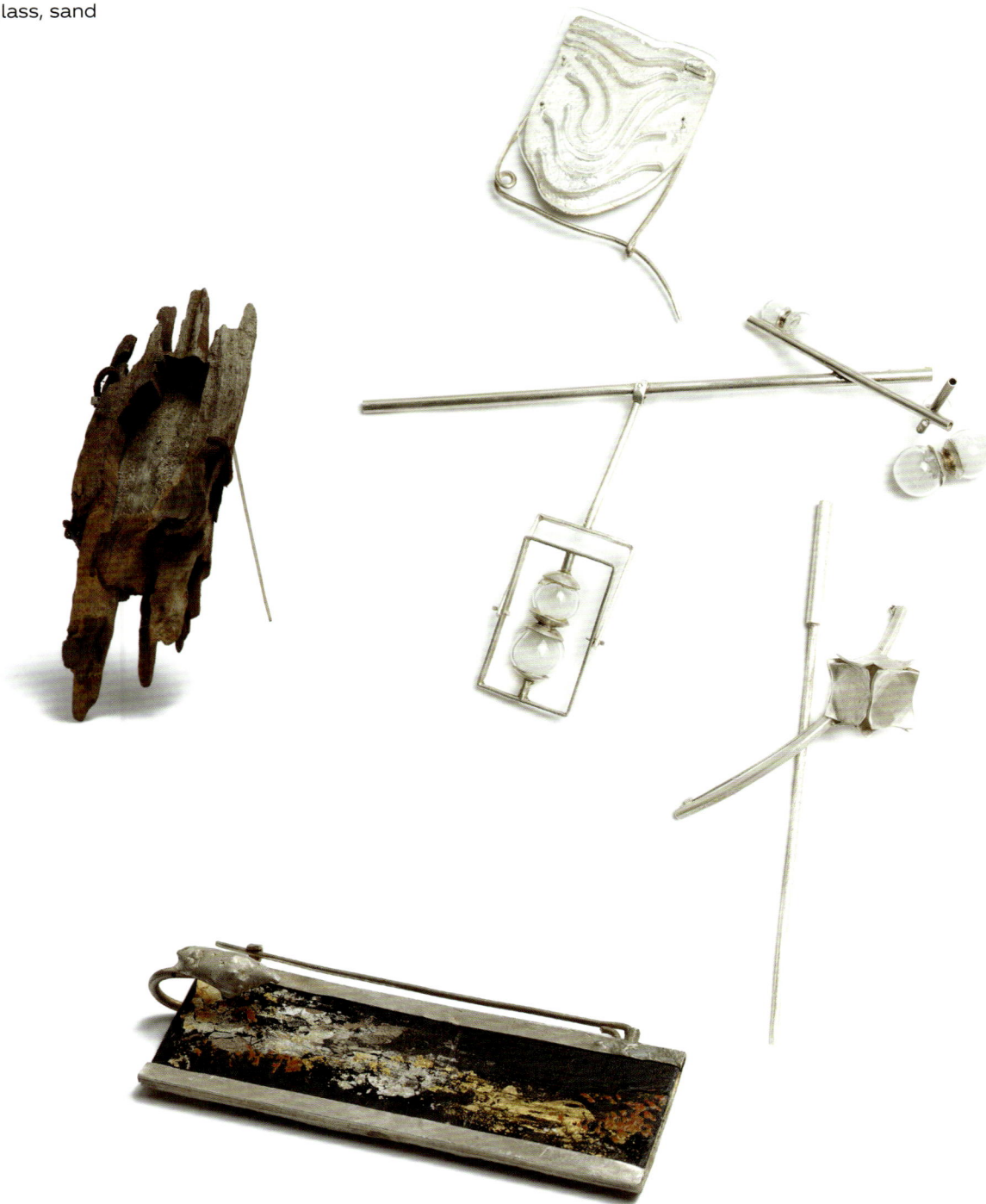

砺行 / Honing

2021 北京国际
首饰艺术展
2021 BEIJING
INTERNATIONAL
JEWELLERY ART
EXHIBITION

134

相由心生
FACE COMES FROM HEART

作者姓名: **韩欣然（中国）**
作品类型: **胸针**
作品材质: **玉石**

Artist: Han Xinran（China）
Type: brooch
Material: jade

砺行 / Honing

2021 北京国际
首饰艺术展
2021 BEIJING
INTERNATIONAL
JEWELLERY ART
EXHIBITION

135

射手座系列 1
SAGITTARIUS SERIES 1

作者姓名: 韩雨蒙（中国）
作品类型: 胸针
作品材质: 18K 黄金，银，航空铝

Artist: Han Yumeng（China）
Type: brooch
Material: 18K yellow gold, silver,
aerospace aluminium

砺行 / Honing

2021 北京国际
首饰艺术展
2021 BEIJING
INTERNATIONAL
JEWELLERY ART
EXHIBITION

136

归于海
RETURNING TO THE SEA

作者姓名：　何凤婷（中国）
作品类型：　项链
作品材质：　银，墨玉，锆石

Artist:　　He Fengting（China）
Type:　　　necklace
Material:　silver, black jade, zircon

隐藏的自然之美 3
HIDDEN NATURE BEAUTY 3

作者姓名：　何萌（中国）
作品类型：　胸针
作品材质：　钛金属（3D 打印）

Artist:　　He Meng（China）
Type:　　　brooch
Material:　titanium (3D print)

砺行 / Honing

2021 北京国际
首饰艺术展
2021 BEIJING
INTERNATIONAL
JEWELLERY ART
EXHIBITION

137

蛮蛮
PRETTY

作者姓名： 何苗（中国）
作品类型： 胸针
作品材质： 925 银，珍珠，玛瑙，蒜皮

Artist: He Miao（China）
Type: brooch
Material: 925 silver, pearl, agate, garlic skin

砺行 / Honing

2021 北京国际
首饰艺术展
2021 BEIJING
INTERNATIONAL
JEWELLERY ART
EXHIBITION

138

作者姓名: 洪书瑶（中国）
作品类型: 胸针
作品材质: 银，木头，锆石

Artist: Hong Shuyao（China）
Type: brooch
Material: silver, wood, zircon

砺行 / Honing

2021 北京国际
首饰艺术展
2021 BEIJING
INTERNATIONAL
JEWELLERY ART
EXHIBITION

139

作者姓名:　胡俊（中国）
作品类型:　胸针
作品材质:　尼龙，银，钢丝

Artist:　　Hu Jun（China）
Type:　　　brooch
Material:　nylon, silver, steel wire

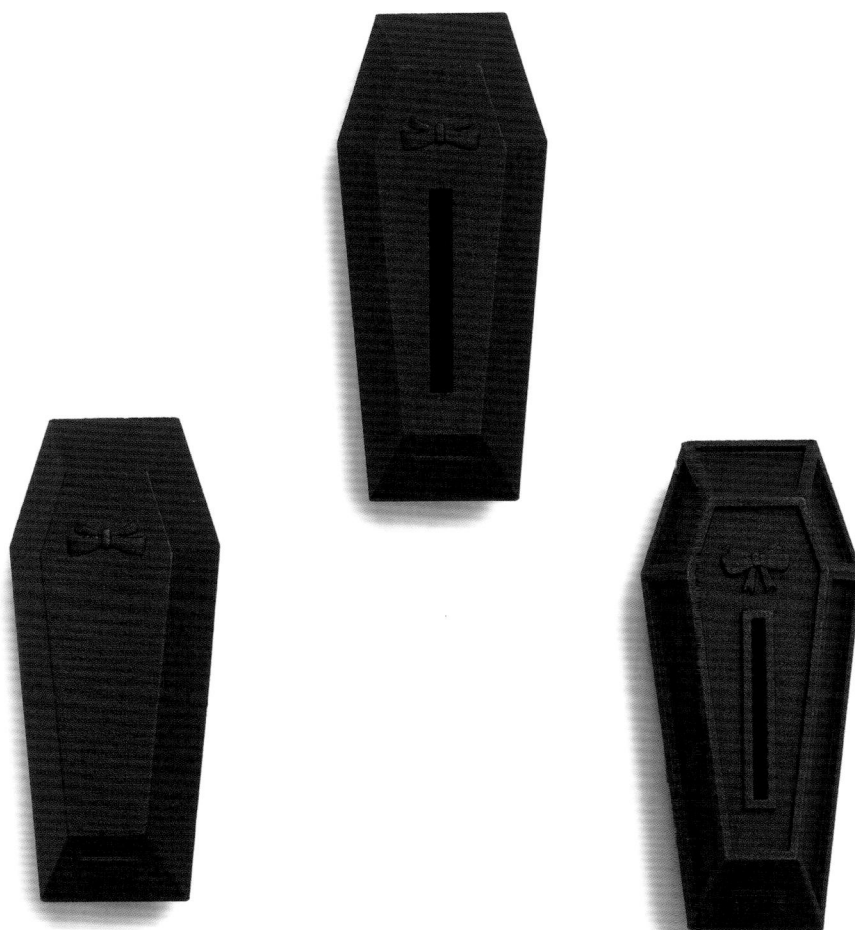

砺行 / Honing

2021 北京国际
首饰艺术展
2021 BEIJING
INTERNATIONAL
JEWELLERY ART
EXHIBITION

140

那时 · 此刻
THEN · NOW

作者姓名： 胡世法（中国）
作品类型： 胸针
作品材质： 银，铝，金箔，照片，橄榄石，钢

Artist: Hu Shifa（China）
Type: brooch
Material: silver, aluminum, gold foil, photo, olivine, steel

未知
UNKNOWN

作者姓名： 胡锶靓（中国）
作品类型： 项链
作品材质： 真丝棉，银，树脂

Artist: Hu Siliang（China）
Type: brooch
Material: silk cotton, silver, resin

砺行 / Honing

2021 北京国际
首饰艺术展
2021 BEIJING
INTERNATIONAL
JEWELLERY ART
EXHIBITION

141

逃离 · 旋转
ESCAPE · SPIN

作者姓名: 黄海欣（中国）
作品类型: 胸针
作品材质: 999 银，925 银，880 银

Artist: Huang Haixin（China）
Type: brooch
Material: 999 silver, 925 silver, 880 silver

新生
NEW BORN

作者姓名: 黄琳（中国）
作品类型: 其他
作品材质: 纤维，拉长石，青金石，水晶

Artist: Huang Lin（China）
Type: other
Material: fiber, elongated stone,
lapis lazuli, crystal

砺行 / Honing

2021 北京国际
首饰艺术展
2021 BEIJING
INTERNATIONAL
JEWELLERY ART
EXHIBITION

142

守的云开见月明——春江花月夜
KEEP THE CLOUDS OPEN
AND SEE THE MOON
— A NIGHT OF FLOWERS AND
MOONLIGHT BY THE SPRING RIVER

作者姓名： 黄思思（中国）
作品类型： 项链
作品材质： 天然真花，树脂，银镀金，珍珠

Artist: Huang Sisi（China）
Type: necklace
Material: natural flowers, resin,
           gold plated silver, pearls

"核"系列一，"核"系列二，
"核"系列三
NUCLEAR SERIES 1, NUCLEAR SERIES 2,
NUCLEAR SERIES 3

作者姓名： 黄煦茜（中国）
作品类型： 项链，胸针
作品材质： 925 银，金箔，黄铜电镀金，银箔，紫水晶，黄水晶

Artist: Huang Xüqian（China）
Type: necklace, brooch
Material: 925 silver, gold leaf, gold planted bass,
           silver foil, amethyst, citrine

砺行 / Honing

2021 北京国际
首饰艺术展
2021 BEIJING
INTERNATIONAL
JEWELLERY ART
EXHIBITION

143

折叠与拼插 II
FOLDING & PIECING II

作者姓名: 黄依瑶（中国）
作品类型: 其他
作品材质: 925 银

Artist: Huang Yiyao（China）
Type: other
Material: 925 silver

Sim 卡卡槽耳环（A-D-JUST）
SIM CARD SLOT EARRINGS (A-D-JUST)

作者姓名: 黄逸（中国）
作品类型: 耳饰
作品材质: 银镀白金

Artist: Huang Yi（China）
Type: earring
Material: silver plated white gold

砺行 / Honing

2021 北京国际
首饰艺术展
2021 BEIJING
INTERNATIONAL
JEWELLERY ART
EXHIBITION

144

"格桑花" 系列首饰——腰带
THE GERBERA COLLECTION JEWELLERY — BELTS

作者姓名: 纪海燕（中国）
作品类型: 其他
作品材质: 925 银，珊瑚，绿松石

Artist: Ji Haiyan（China）
Type: other
Material: 925 silver, coral and turquoise

旋
CYCLONE

作者姓名: 江秉黎（中国）
作品类型: 胸针
作品材质: 珍珠，合成立方氧化锆，银镀 18K 金

Artist: Jiang Bingli（China）
Type: brooch
Material: pearl, synthetic cubic zirconia,
18K gold plated silver

砺行 / Honing

2021 北京国际
首饰艺术展
2021 BEIJING
INTERNATIONAL
JEWELLERY ART
EXHIBITION

145

作者姓名: 江婷（中国）
作品类型: 项链
作品材质: 925 银，"bullseye" 玻璃

Artist: Jiang Ting（China）
Type: necklace
Material: 925 silver, "bullseye" glass

砺行 / Honing

2021 北京国际
首饰艺术展
2021 BEIJING
INTERNATIONAL
JEWELLERY ART
EXHIBITION

146

发声
SPEAKING

作者姓名:　姜倩（中国）
作品类型:　胸针
作品材质:　999 银

Artist:　　Jiang Qian（China)
Type:　　　brooch
Material:　999 silver

砺行 / Honing

2021 北京国际
首饰艺术展
2021 BEIJING
INTERNATIONAL
JEWELLERY ART
EXHIBITION

147

离开
GET AWAY

作者姓名: 蒋旻（中国）
作品类型: 其他
作品材质: 黄铜，草木灰，泡棉材料

Artist: Jiang Min（China）
Type: other
Material: brass, plant ash, foam material

静候 1
QUIET WAIT 1

作者姓名: 金翠玲（中国）
作品类型: 胸针
作品材质: 925 银，宣纸

Artist: Jin Cuiling（China）
Type: brooch
Material: 925 silver, rice paper

砺行 / Honing

2021 北京国际
首饰艺术展
2021 BEIJING
INTERNATIONAL
JEWELLERY ART
EXHIBITION

148

作者姓名: **金若雨（中国）**
作品类型: **其他**
作品材质: **白玉，玛瑙，金箔，银，树脂，绸缎**
Artist: **Jin Ruoyu（China）**
Type: **other**
Material: **white jade, agate, gold foil, silver, resin, silk**

砺行 / Honing

2021北京国际
首饰艺术展
2021 BEIJING
INTERNATIONAL
JEWELLERY ART
EXHIBITION

149

未知
UNKNOWN

作者姓名: 金晓蕊（中国）
作品类型: 胸针
作品材质: 钛金属丝，硅胶，玻璃珠

Artist: Jin Xiaorui（China）
Type: brooch
Material: titanium wire, silicone, glass beads

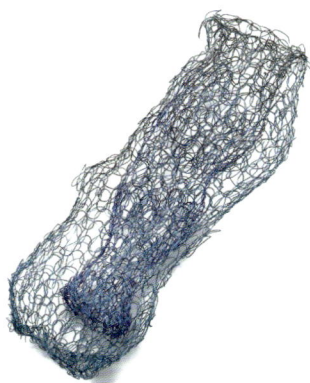

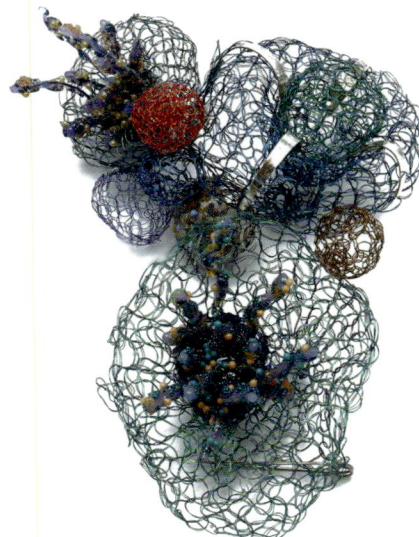

心镜
HEART MIRROR

作者姓名: 金雪凌（中国）
作品类型: 项链
作品材质: 24K 黄金，黄铜，亚克力

Artist: Jin Xueling（China）
Type: necklace
Material: 24K yellow gold, brass, acrylic

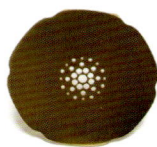

砺行 / Honing

2021 北京国际
首饰艺术展
2021 BEIJING
INTERNATIONAL
JEWELLERY ART
EXHIBITION

150

监护人
GUARDIAN

作者姓名: 金知瑞（中国）
作品类型: 其他
作品材质: 麻布，大漆，银粉，金粉

Artist: Jin Zhirui（China）
Type: other
Material: sackcloth, lacquer,
silver powder and gold powder

任尔东西南北风 1，任尔东西南北风 2
FROM WHICHEVER DIRECTION THE WINDS LEAD 1,
FROM WHICHEVER DIRECTION THE WINDS LEAD 2

作者姓名: 晋文捷（中国）
作品类型: 胸针，戒指
作品材质: 银镀金，树脂

Artist: Jin Wenjie（China）
Type: brooch, ring
Material: gold plated silver, resin

砺行 / Honing

2021 北京国际
首饰艺术展
2021 BEIJING
INTERNATIONAL
JEWELLERY ART
EXHIBITION

151

枯木、丰碑、摇篮
DEADWOOD, MONUMENT, CRADLE

作者姓名： 康铂雍（中国）
作品类型： 戒指
作品材质： 银，钢筋，水泥

Artist: Kang Boyong（China）
Type: ring
Material: silver, steel, concrete

蜕变
TRANSFORMATION

作者姓名： 李佰羲（中国）
作品类型： 项链
作品材质： 纸，蚕茧，银丝，珍珠

Artist: Li Baixi（China）
Type: necklace
Material: paper, cocoons, silver silk, pearls

砺行 / Honing

2021 北京国际
首饰艺术展
2021 BEIJING
INTERNATIONAL
JEWELLERY ART
EXHIBITION

152

西江月
WEST RIVER MOON

作者姓名: 李冰剑（中国）
作品类型: 胸针
作品材质: 18K 金，翡翠

Artist: Li Bingjian（China）
Type: brooch
Material: 18K gold, jadeite

刘关张
LIU BEI, GUAN YU AND ZHANG FEI

作者姓名: 李登登（中国）
作品类型: 胸针
作品材质: 银，现代珐琅

Artist: Li Dengdeng（China）
Type: brooch
Material: silver, modern enamel

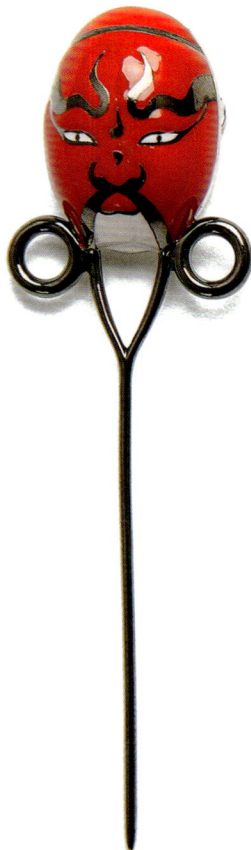

砺行 / Honing

2021 北京国际
首饰艺术展
2021 BEIJING
INTERNATIONAL
JEWELLERY ART
EXHIBITION

153

花笠
ECHEVERIA

作者姓名: 李靖仪（中国）
作品类型: 项链，戒指，手镯
作品材质: 925 银，淡水珍珠，亚克力

Artist: Li Jingyi（China）
Type: necklace, ring, bracelet
Material: 925 silver, fresh water pearls,
acrylic

砥行 / Honing

2021北京国际
首饰艺术展
2021 BEIJING
INTERNATIONAL
JEWELLERY ART
EXHIBITION

154

枯石
DEAD STONE

作者姓名：　**李萌（中国）**
作品类型：　**耳饰**
作品材质：　**925 银，金泊，锆石**

Artist:　　**Li Meng（China）**
Type:　　　**earring**
Material:　　**925 silver, gold leaf, zircon**

2021 北京国际
首饰艺术展
2021 BEIJING
INTERNATIONAL
JEWELLERY ART
EXHIBITION

155

墨花吟系列——1
FLOWER SHADOW CHANT COLLECTION — 1

作者姓名： 李楠（中国）
作品类型： 胸针
作品材质： 亚克力，924 银，14K 黄金链子，不锈钢针

Artist: Li Nan（China）
Type: brooch
Material: acrylic, 924 silver, 14K yellow gold chains,
stainless steel needles

何处归
WHERE TO

作者姓名： 李珊珊（中国）
作品类型： 胸针
作品材质： 檀木，树脂，银，水晶

Artist: Li Shanshan（China）
Type: brooch
Material: sandalwood, resin, silver, crystal

砺行 / Honing

2021 北京国际
首饰艺术展
2021 BEIJING
INTERNATIONAL
JEWELLERY ART
EXHIBITION

156

石情
STONE MOOD

作者姓名:	李亭雨（中国）
作品类型:	胸针
作品材质:	冷珐琅，银，珍珠
Artist:	Li Tingyu（China）
Type:	brooch
Material:	cold enamel, silver, pearls

安博利特一号
AMBILIGHT NO. 1

作者姓名:	李文斯（中国）
作品类型:	耳饰，项链
作品材质:	钛金属， 天然彩色钻石
Artist:	Li Wensi（China）
Type:	earring, necklace
Material:	titanium, natural color diamond

砺行 / Honing

2021 北京国际
首饰艺术展
2021 BEIJING
INTERNATIONAL
JEWELLERY ART
EXHIBITION

157

新生
NEW BORN

作者姓名: 李汶轩（中国）
作品类型: 胸针
作品材质: 925 银，岫玉

Artist: Li Wenxuan（China）
Type: brooch
Material: 925 silver, jade

首饰，容器
JEWELLERY, THE CONTAINER

作者姓名: 李雅婷（中国）
作品类型: 项链
作品材质: 银，银镀金

Artist: Li Yating（China）
Type: neclace
Material: silver, gold plated silver

砺行 / Honing

2021 北京国际
首饰艺术展
2021 BEIJING
INTERNATIONAL
JEWELLERY ART
EXHIBITION

158

束缚 1
BOUND 1

作者姓名:	李岩（中国）
作品类型:	胸针
作品材质:	黄铜，金属网
Artist:	Li Yan（China）
Type:	brooch
Material:	brass, metal mesh

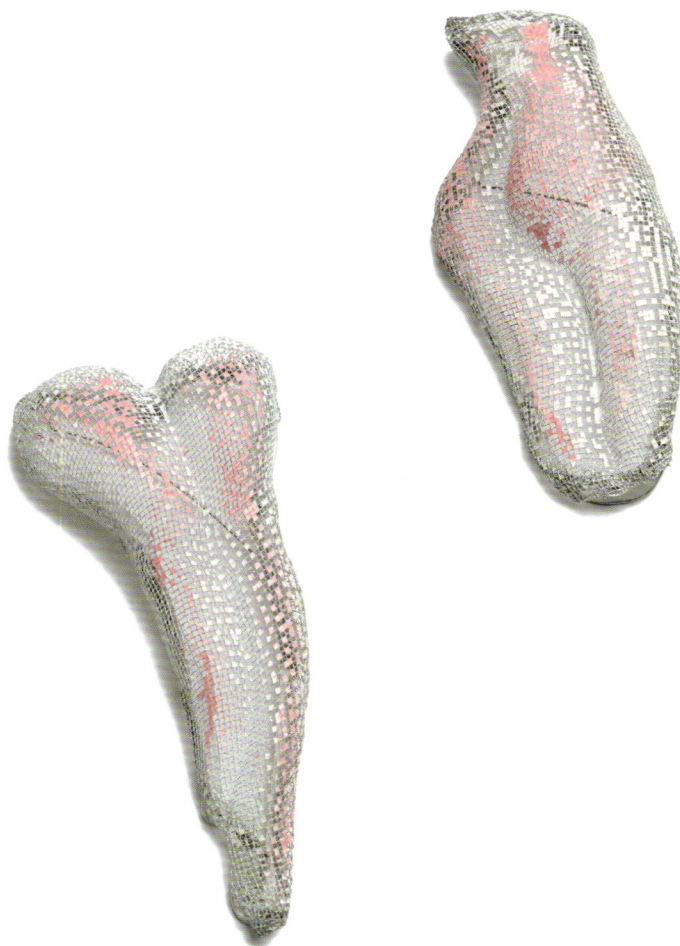

再青春
AGAIN THE YOUTH

作者姓名:	李阳（中国）
作品类型:	项链
作品材质:	树脂，漆，银镀金
Artist:	Li Yang（China）
Type:	necklace
Material:	resin, lacquer, gold plated silver

砺行 / Honing

2021 北京国际
首饰艺术展
2021 BEIJING
INTERNATIONAL
JEWELLERY ART
EXHIBITION

159

柒时
SEVEN O'CLOCK

作者姓名: 李怡（中国）
作品类型: 胸针
作品材质: 黑牛角，银

Artist: Li Yi（China）
Type: brooch
Material: black horn, silver

绽放
BLOOM

作者姓名: 李梓峰（中国）
作品类型: 胸针
作品材质: 18K 黄金，猫眼石，钻石

Artist: Li Zifeng（China）
Type: brooch
Material: 18K yellow gold, opals, diamonds

砺行 / Honing

2021 北京国际
首饰艺术展
2021 BEIJING
INTERNATIONAL
JEWELLERY ART
EXHIBITION

160

蝶
BUTTERFLY

作者姓名： **李宗跃（中国）**
作品类型： **胸针**
作品材质： **银镀金，翡翠，氧化锆**

Artist: **Li Zongyue（China）**
Type: **brooch**
Material: **gold plated silver, jade, zirconia**

砺行 / Honing

2021 北京国际
首饰艺术展
2021 BEIJING
INTERNATIONAL
JEWELLERY ART
EXHIBITION

161

透明诗饰系列 I
ONE OF THE TRANSPARENT POEM DECORATION SERIES I

作者姓名: 林恩迎（中国）
作品类型: 项链
作品材质: PE 塑料，水晶线

Artist: Lin Enying（China）
Type: necklace
Material: PE plastic, crystal wire

砺行 / Honing

2021 北京国际
首饰艺术展
2021 BEIJING
INTERNATIONAL
JEWELLERY ART
EXHIBITION

162

泉舞乐韵，玉见——未来，云兴水逸
SPRING DANCE MUSIC, SEE YOU — THE FUTURE, FREEDOM LIKE THE CLOUD LUCK LIKE THE WATER

作者姓名： 林弘裕（中国）

作品类型： 胸针

作品材质： 天然翡翠，黑玛瑙（又称安立士石），白钻，18K 铂金，
黑玛瑙，红宝石，红玛瑙，青金石，贝母

Artist: Lin Hongyu（China）

Type: brooch

Material: natural jade, black agate, white diamond,
18K platinum, ruby, red agate, lapis lazuli, fritillaria

听行 / Honing

2021 北京国际
首饰艺术展
2021 BEIJING
INTERNATIONAL
JEWELLERY ART
EXHIBITION

163

石出
OUT OF THE STONE

作者姓名: 刘建钊（中国）
作品类型: 其他
作品材质: 白玉

Artist: Liu Jianzhao（China）
Type: other
Material: white jade

砺行 / Honing

2021 北京国际
首饰艺术展
2021 BEIJING
INTERNATIONAL
JEWELLERY ART
EXHIBITION

164

共生
THE SYMBIOTIC

作者姓名： 刘娇（中国）
作品类型： 胸针
作品材质： 银，珍珠，海绵，亚克力

Artist: Liu Jiao（China）
Type: brooch
Material: silver, pearl, sponge, acrylic

序源
SEQUENCE SOURCE

作者姓名： 刘劲（中国）
作品类型： 项链
作品材质： 银，丝线，布

Artist: Liu Jing（China）
Type: necklace
Material: silver, silk thread, cloth

师行 / Honing

2021 北京国际
首饰艺术展
2021 BEIJING
INTERNATIONAL
JEWELLERY ART
EXHIBITION

165

萌生
INITIATION

作者姓名: 刘磊（中国）
作品类型: 戒指
作品材质: 珊瑚，珍珠，金箔，银

Artist: Liu Lei（China）
Type: ring
Material: coral, pearl, gold leaf, silver

方少时系列——拉哨
JUST AS THE YOUNG — PULL THE WHISTLE

作者姓名: 刘立程（中国）
作品类型: 项链
作品材质: 银，红绳

Artist: Liu Licheng（China）
Type: necklace
Material: silver, red thread

砺行 / Honing

2021 北京国际
首饰艺术展
2021 BEIJING
INTERNATIONAL
JEWELLERY ART
EXHIBITION

166

福袋 2020：神农本草经
LUCKY BAG 2020 :
SHENNONG'S HERBAL CLASSIC OF MATERIA MEDICA

作者姓名： 刘骁（中国）
作品类型： 其他
作品材质： 医用外科口罩，《神农本草经》内文，桑蚕丝

Artist: Liu Xiao（China）
Type: other
Material: medical surgical masks,
Shennong Materia Medica text, mulberry silk

2021 北京国际
首饰艺术展
2021 BEIJING
INTERNATIONAL
JEWELLERY ART
EXHIBITION

167

重生
REBIRTH

作者姓名：　刘雪茜（中国）
作品类型：　项链
作品材质：　3D 打印废弃树脂，金属

Artist:　　Liu Xueqian（China）
Type:　　　necklace
Material:　 3D printing of waste resin, metal

观木
CONCEPT OF WOOD

作者姓名：　刘泽慧（中国）
作品类型：　项链
作品材质：　黄铜，木头，14K 金

Artist:　　Liu Zehui（China）
Type:　　　necklace
Material:　 brass, wood, 14K gold

砺行 / Honing

2021 北京国际
首饰艺术展
2021 BEIJING
INTERNATIONAL
JEWELLERY ART
EXHIBITION

168

瑶之徙之觅火
YAO'S JOURNEY IN SEARCH OF FIRE

作者姓名: 卢念念（中国）
作品类型: 项链
作品材质: 银，高温珐琅

Artist: Lu Niannian（China）
Type: necklace
Material: silver, high temperature enamel

梦幻的边缘
EDGE OF DREAMS

作者姓名: 卢艺（中国）
作品类型: 胸针，戒指
作品材质: 钛，银，珐琅，不锈钢，珍珠

Artist: Lu Yi（China）
Type: brooch, ring
Material: titanium, silver, enamel,
stainless steel, pearl

砺行 / Honing

2021 北京国际
首饰艺术展
2021 BEIJING
INTERNATIONAL
JEWELLERY ART
EXHIBITION

169

望
LOOKING

作者姓名: 卢雨梦（中国）
作品类型: 胸针
作品材质: 陶瓷，手绘青花，925 银电镀金

Artist: Lu Yumeng（China）
Type: brooch
Material: ceramic, hand-painted blue and white porcelain, 925 gold plated silver

隙光清影——绕
LIGHT AND CLEAR SHADOWS
FROM THE CREVICES — WINDING

作者姓名: 鲁霁萱（中国）
作品类型: 项链
作品材质: 银黏土，银，铜，树脂，大漆，珐琅

Artist: Lu Jixuan（China）
Type: necklace
Material: silver clay, silver, copper, resin, lacquer, enamel

砺行 / Honing

2021 北京国际
首饰艺术展
2021 BEIJING
INTERNATIONAL
JEWELLERY ART
EXHIBITION

170

携手——联结
COOPERATION & UNION

作者姓名: 鲁硕（中国）
作品类型: 项链
作品材质: 白玉，银镀金

Artist: Lu Shuo（China）
Type: necklace
Material: white jade, gold plated silver

砺行 / Honing

2021 北京国际
首饰艺术展
2021 BEIJING
INTERNATIONAL
JEWELLERY ART
EXHIBITION

171

引
GUIDING

作者姓名: 　陆清鹇（中国）
作品类型: 　项链
作品材质: 　银

Artist: 　Lu Qingyuan（China）
Type: 　necklace
Material: 　silver

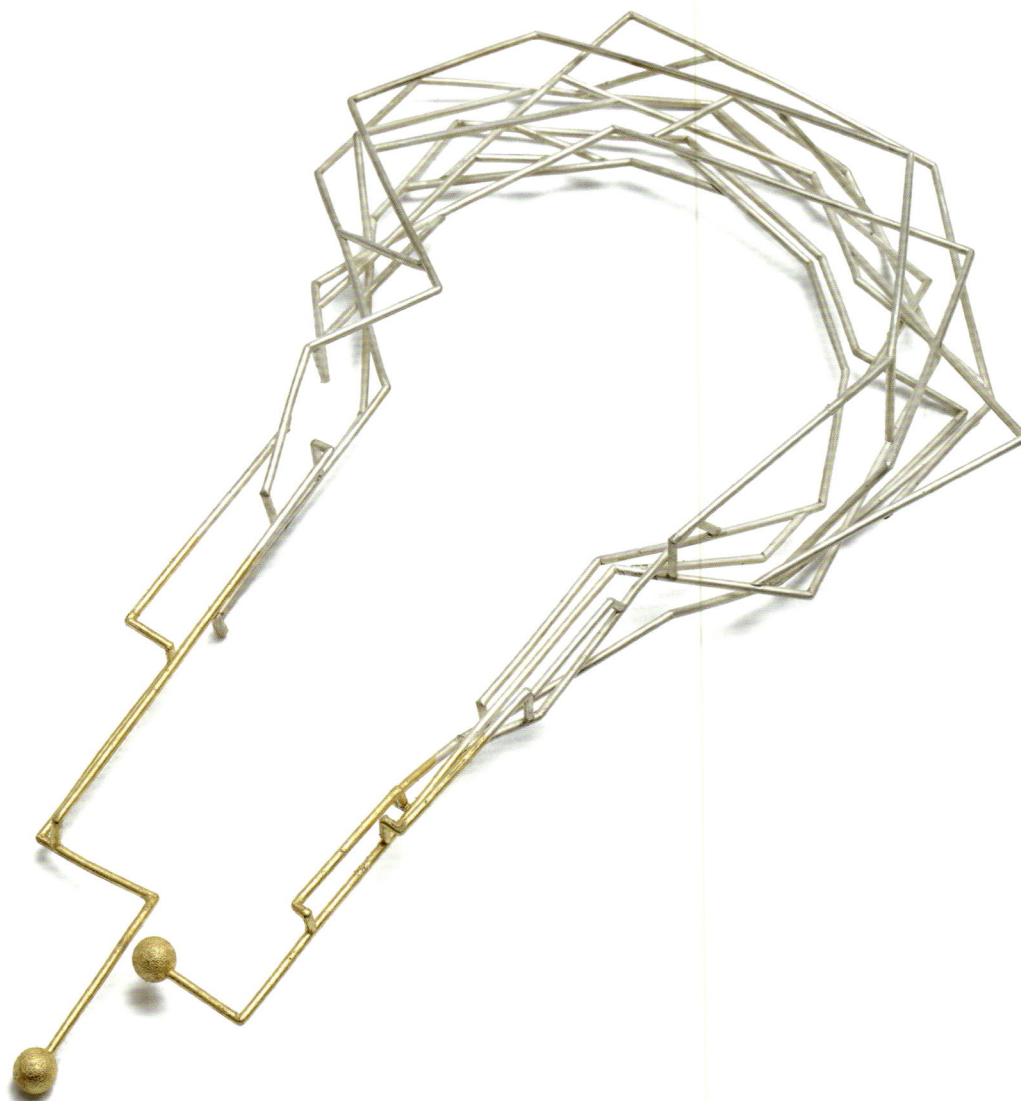

砺行 / Honing

2021 北京国际
首饰艺术展
2021 BEIJING
INTERNATIONAL
JEWELLERY ART
EXHIBITION

172

勺与时光
SPOON WITH TIME

作者姓名: 路小五（中国）
作品类型: 项链
作品材质: 银

Artist: **Lu Xiaowu（China）**
Type: **necklace**
Material: **silver**

砺行 / Honing

2021 北京国际
首饰艺术展
2021 BEIJING
INTERNATIONAL
JEWELLERY ART
EXHIBITION

173

春逝水
LOST SPRING

作者姓名： 栾建霞（中国）
作品类型： 戒指，手镯
作品材质： 银，祖母绿，缅甸尖晶，立方氧化锆

Artist: Luan Jianxia（China）
Type: ring, bracelet
Material: silver, emerald, Myanmar spinel,
cubic zirconia

我在你身上看见了极光
I SEE AURORA ON YOUR BODY

作者姓名： 罗鸿蒙（中国）
作品类型： 项链
作品材质： 阳极氧化钛，氧化银

Artist: Luo Hongmeng（China）
Type: necklace
Material: anodic titanium oxide, silver oxide

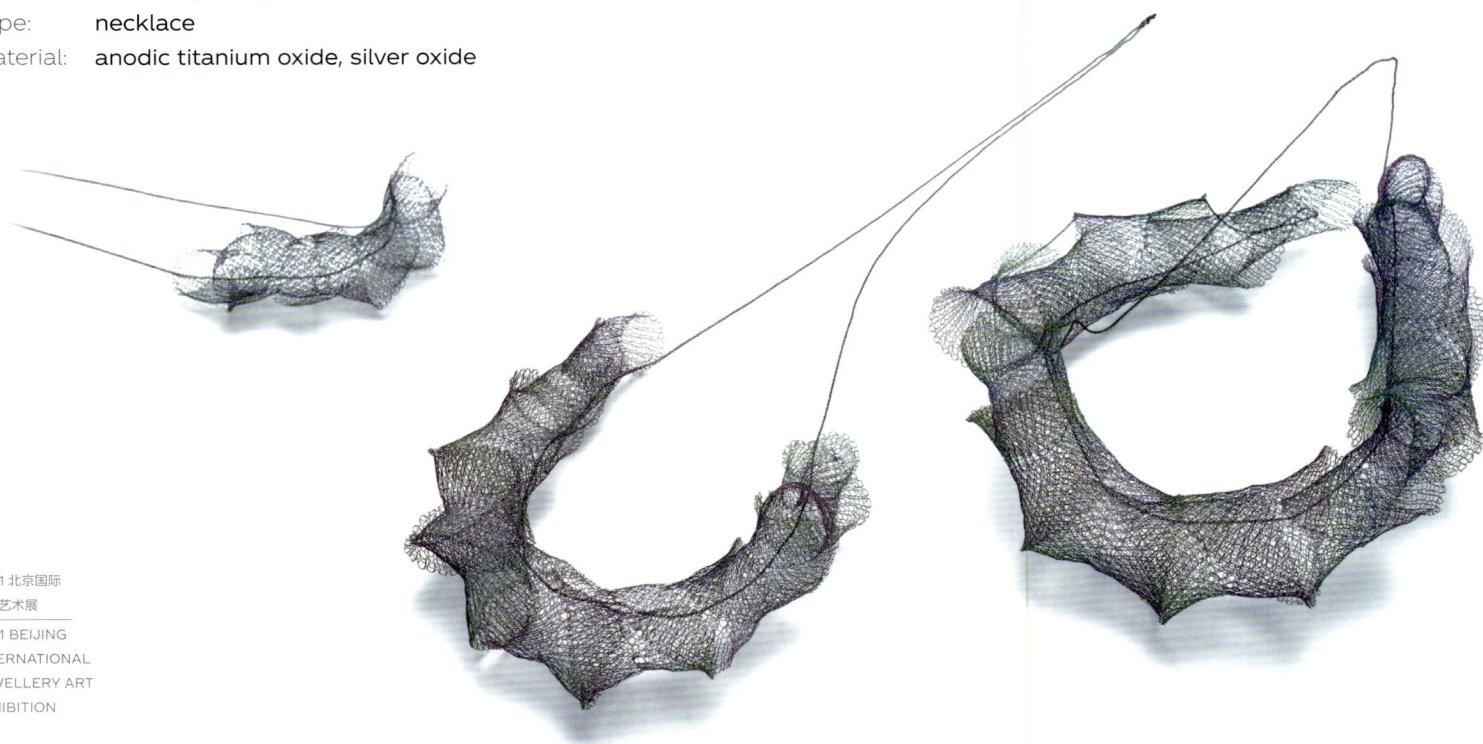

砺行 / Honing

2021 北京国际
首饰艺术展
2021 BEIJING
INTERNATIONAL
JEWELLERY ART
EXHIBITION

174

家的味道
THE TASTE OF HOME

作者姓名: 罗晶晶（中国）
作品类型: 戒指
作品材质: 银

Artist: Luo Jingjing（China）
Type: ring
Material: silver

云外惊飞
FLY ABOVE THE CLOUDS

作者姓名: 罗理婷（中国）
作品类型: 胸针
作品材质: 黑檀，银，锆石

Artist: Luo Liting（China）
Type: brooch
Material: ebony, silver, zircon

砺行 / Honing

2021 北京国际
首饰艺术展
2021 BEIJING
INTERNATIONAL
JEWELLERY ART
EXHIBITION

175

你好！父母
HELLO! PARENTS

作者姓名：　吕雪原（中国）
作品类型：　其他
作品材质：　黄铜，羽毛

Artist:　　Lü Xueyuan（China）
Type:　　　other
Material:　brass, feather

无意识的喜悦
UNCONSCIOUS JOY

作者姓名：　吕赟蛟（中国）
作品类型：　胸针
作品材质：　银，七宝

Artist:　　Lü Yunjiao（China）
Type:　　　brooch
Material:　silver, seven traditional materials

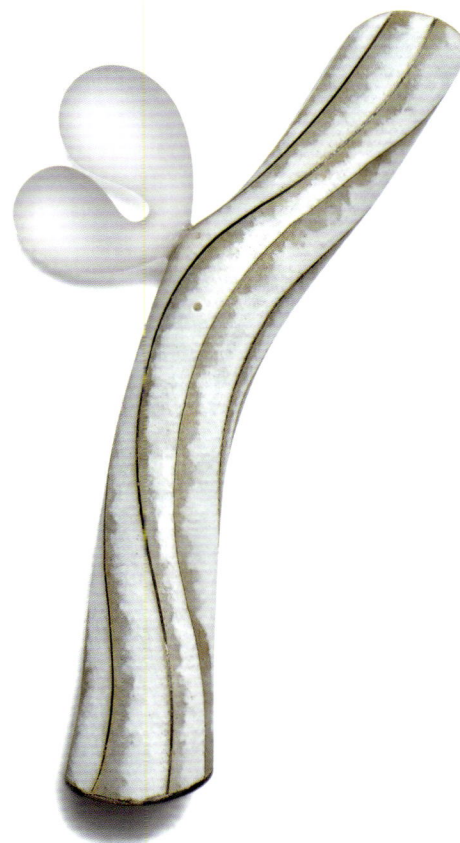

砺行 / Honing

2021 北京国际
首饰艺术展
2021 BEIJING
INTERNATIONAL
JEWELLERY ART
EXHIBITION

176

娲蛙哇
WA HUO WA WA

作者姓名: 马瑞希（中国）
作品类型: 胸针
作品材质: 阳极氧化钛，黄铜

Artist: Ma Ruixi（China）
Type: necklace
Material: anodic titanium oxide, brass

源系列 2
SOURCE SERIES 2

作者姓名: 马诗彧（中国）
作品类型: 胸针
作品材质: 银，紫铜，黄铜

Artist: Ma Shiyu（China）
Type: brooch
Material: silver, copper, brass

砺行 / Honing

2021 北京国际
首饰艺术展
2021 BEIJING
INTERNATIONAL
JEWELLERY ART
EXHIBITION

177

莲 · 藻 1
LOTUS·ALGAE 1

作者姓名: 马思娴（中国）
作品类型: 胸针
作品材质: 银

Artist: Ma Sixian（China）
Type: brooch
Material: silver

雍和之境 1
MIND OF YONGHE 1

作者姓名: 马宇婷（中国）
作品类型: 胸针
作品材质: 树脂，黄铜，合金

Artist: Ma Yuting（China）
Type: brooch
Material: resin, brass, alloy

砥行 / Honing

2021 北京国际
首饰艺术展
2021 BEIJING
INTERNATIONAL
JEWELLERY ART
EXHIBITION

178

"过番"
TO MAKE A LIVING

作者姓名: 麦扬（中国）
作品类型: 戒指
作品材质: 不锈钢，黄铜

Artist: Mai Yang（China）
Type: ring
Material: stainless steel, brass

大渡桥横铁索寒
COLD THE IRON-CHAIN SPANS OF TATU'S BRIDGE

作者姓名: 孟亮（中国）
作品类型: 胸针
作品材质: 墨翠，黑檀木，银

Artist: Meng Liang（China）
Type: brooch
Material: emerald, ebony, silver

砺行 / Honing

2021 北京国际
首饰艺术展
2021 BEIJING
INTERNATIONAL
JEWELLERY ART
EXHIBITION

179

梦魇
NIGHTMARE

作者姓名: 聂可实（中国）
作品类型: 其他
作品材质: 玻璃，火山石

Artist: Nie Keshi（China）
Type: other
Material: glass, volcanic stone

椎的支撑
SUPPORT OF SPINE

作者姓名: 潘璠（中国）
作品类型: 胸针
作品材质: 银

Artist: Pan Fan（China）
Type: brooch
Material: silver

砺行 / Honing

2021 北京国际
首饰艺术展
2021 BEIJING
INTERNATIONAL
JEWELLERY ART
EXHIBITION

180

勋章
MEDAL

作者姓名: 潘奕吉（中国）
作品类型: 胸针
作品材质: 牛角废料，回收银

Artist: Pan Yiji（China）
Type: brooch
Material: horn waste, recycling silver

永恒
INFINITE

作者姓名: 彭星星（中国）
作品类型: 项链
作品材质: 钛金属，银

Artist: Peng Xingxing（China）
Type: necklace
Material: titanium, silver

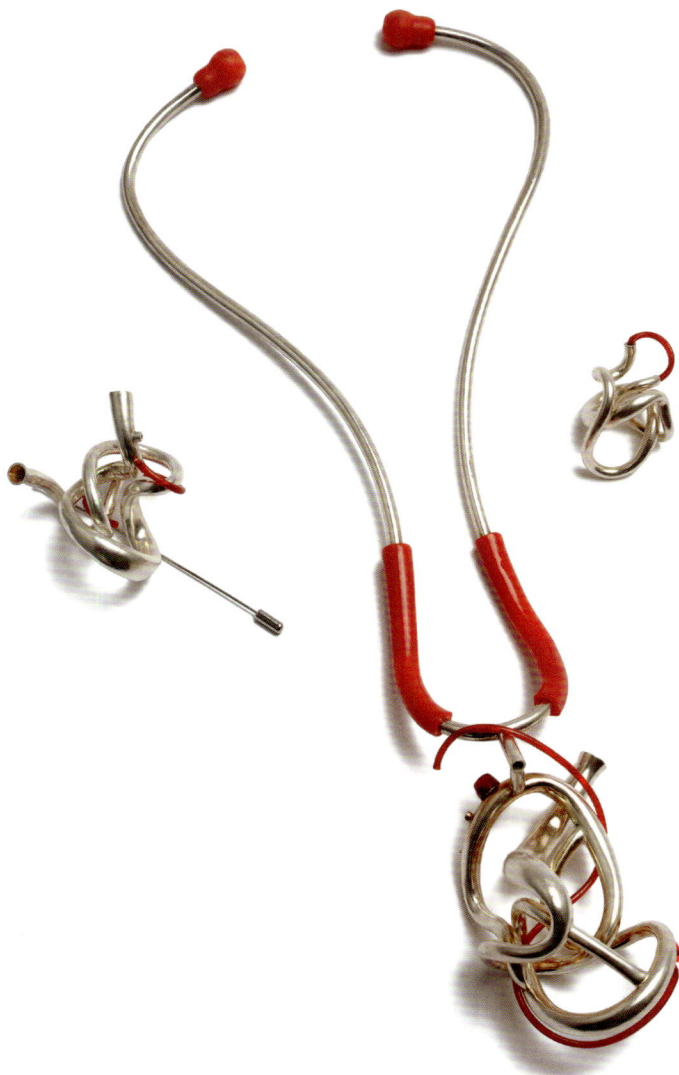

2021 北京国际
首饰艺术展
2021 BEIJING
INTERNATIONAL
JEWELLERY ART
EXHIBITION

181

自得
COMPLACENT

作者姓名：　祁子芮（中国）
作品类型：　胸针
作品材质：　18K 黄金，珊瑚，翡翠，钻石

Artist:　　Qi Zirui（China）
Type:　　　brooch
Material:　18K yellow gold, coral, jade, diamonds

百虫记——蚂蚁
INSECTS — ANTS

作者姓名：　钱丽霞（中国）
作品类型：　胸针
作品材质：　925 银，蜜蜡，星光蓝宝石，钢丝

Artist:　　Qian Lixia（China）
Type:　　　brooch
Material:　925 silver, beeswax, star sapphire, steel wire

砺行 / Honing

2021 北京国际
首饰艺术展
2021 BEIJING
INTERNATIONAL
JEWELLERY ART
EXHIBITION

182

生命之花 ——
水知道答案 No1. 3 fold
THE FLOWER OF LIFE —
WATER KNOWS THE ANSWER NO1.3 FOLD

作者姓名: 任聪（中国）
作品类型: 胸针
作品材质: 珐琅，999 银，925 银，不锈钢

Artist: Ren Cong（China）
Type: brooch
Material: enamel, 999 silver,
925 silver, stainless steel

暗示
HINT

作者姓名: 任海明（中国）
作品类型: 戒指
作品材质: 黄铜，银，钛

Artist: Ren Haiming（China）
Type: ring
Material: brass, silver, titanium

砺行 / Honing

2021 北京国际
首饰艺术展
2021 BEIJING
INTERNATIONAL
JEWELLERY ART
EXHIBITION

183

蝴蝶兰
PHALAENOPSIS

作者姓名:　**任俊颖（中国）**
作品类型:　**胸针**
作品材质:　**银**

Artist:　　**Ren Junying（China）**
Type:　　　**brooch**
Material:　**silver**

砺行 / Honing

2021 北京国际
首饰艺术展
2021 BEIJING
INTERNATIONAL
JEWELLERY ART
EXHIBITION

184

新生
NEWBORN

作者姓名：	邵翊恩（中国）
作品类型：	胸针
作品材质：	925 银，高温珐琅，手绘珐琅
Artist:	Shao Hongen（China）
Type:	brooch
Material:	925 silver, high temperature enamel, hand painted enamel

春涧
THE SPRING WATERS

作者姓名：	沈秀文，何彦欣（中国）
作品类型：	项链
作品材质：	银，合成蓝宝
Artist:	Shen Xiuwen, He Yanxin（China）
Type:	necklace
Material:	silver, synthetic sapphire

砺行 / Honing

2021 北京国际
首饰艺术展
2021 BEIJING
INTERNATIONAL
JEWELLERY ART
EXHIBITION

185

脆弱的力量
FRAGILE POWER

作者姓名: 石芮宁（中国）
作品类型: 胸针
作品材质: 925 银

Artist: Shi Ruining（China）
Type: brooch
Material: 925 silver

来自大海 A
FROM THE SEA A

作者姓名: 时俊（中国）
作品类型: 胸针
作品材质: 鹅卵石，软陶，贝壳，水晶

Artist: Shi Jun（China）
Type: brooch
Material: cobblestones, cotta, shells, crystals

砺行 / Honing

2021 北京国际
首饰艺术展
2021 BEIJING
INTERNATIONAL
JEWELLERY ART
EXHIBITION

186

气孔 Ⅲ
STOMATAL Ⅲ

作者姓名: 宋培铭（中国）
作品类型: 胸针
作品材质: 银镀 18K 黄金，3D 氧化铝陶瓷

Artist: Song Peiming（China）
Type: brooch
Material: 18K yellow gold plated silver, 3D alumina ceramics

玉米
CORN

作者姓名: 宋鑫子（中国）
作品类型: 耳饰
作品材质: 硅胶，染色剂，925 银

Artist: Song Xinzi（China）
Type: earring
Material: silicone, stain, 925 silver

砺行 / Honing

2021 北京国际
首饰艺术展
2021 BEIJING
INTERNATIONAL
JEWELLERY ART
EXHIBITION

187

数字噪音：基于曲线方程的首饰数字化生成设计
MATH CHAOS: BASE ON CURVE FORMULA JEWELER DIGITAL GENERATED DESIGN

作者姓名：　宋懿（中国）
作品类型：　耳饰，头饰
作品材质：　尼龙，银

Artist:　　Song Yi（China）
Type:　　earring, headwear
Material:　nylon, silver

砺行 / Honing

2021 北京国际
首饰艺术展
2021 BEIJING
INTERNATIONAL
JEWELLERY ART
EXHIBITION

188

一颗
ONE

作者姓名： 苏艺珂（中国）
作品类型： 胸针
作品材质： 陶瓷，925 银，雕塑膏

Artist: Su Yike（China）
Type: brooch
Material: ceramics, 925 silver, sculpture paste

危 · 福
DANGER · BLESSING

作者姓名： 苏芸菲（中国）
作品类型： 胸针
作品材质： 925 银，铜，钛，红酸枝木，
　　　　　 亚克力，大漆，综合材料

Artist: Su Yunfei（China）
Type: brooch
Material: 925 silver, copper, titanium,
red acid wood, acrylic, lacquer,
composite materials

砺行 / Honing

2021 北京国际
首饰艺术展
2021 BEIJING
INTERNATIONAL
JEWELLERY ART
EXHIBITION

189

作者姓名: 隋建国（中国）
作品类型: 摆件
作品材质: 玉，金，钻石

Artist: Sui Jianguo（China）
Type: object
Material: jade, gold, diamonds

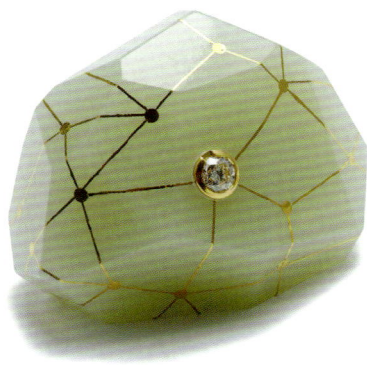

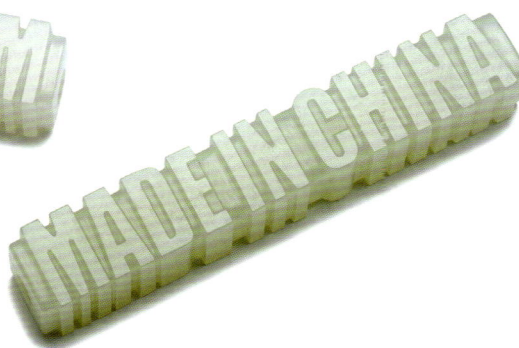

砺行 / Honing

2021 北京国际
首饰艺术展
2021 BEIJING
INTERNATIONAL
JEWELLERY ART
EXHIBITION

190

心流·观想
FLOW·VISUALIZ

作者姓名:　孙梦影（中国）
作品类型:　戒指
作品材质:　银，磁铁，光纤，铁，PVC 管

Artist:　　**Sun Mengying（China）**
Type:　　　**ring**
Material:　**silver, magnets, fiber optics, iron, PVC pipe**

砺行 / Honing

2021 北京国际
首饰艺术展

2021 BEIJING
INTERNATIONAL
JEWELLERY ART
EXHIBITION

191

安题：L.1
ANTI: L.1

作者姓名： 孙铭睿（中国）
作品类型： 胸针
作品材质： 银

Artist: **Sun Mingrui（China）**
Type: **brooch**
Material: **silver**

砺行 / Honing

2021 北京国际
首饰艺术展
2021 BEIJING
INTERNATIONAL
JEWELLERY ART
EXHIBITION

192

十二花神系列
12 FLOWER GODS' SERIES

作者姓名： 孙平（中国）
作品类型： 其他
作品材质： 金，银，红珊瑚，翡翠等

Artist: Sun Ping（China）
Type: other
Material: gold, silver, red coral, jade etc.

师行 / Honing

2021 北京国际
首饰艺术展
2021 BEIJING
INTERNATIONAL
JEWELLERY ART
EXHIBITION

193

殖──记忆
COLONIZATION — MEMORY

作者姓名: 孙秋爽（中国）
作品类型: 项链
作品材质: 银，冷珐琅

Artist: Sun Qiushuang（China）
Type: necklace
Material: silver, cold enamel

蜕
METAMORPHOSIS

作者姓名: 孙士尧（中国）
作品类型: 项链
作品材质: 纸，颜料，线

Artist: Sun Shiyao（China）
Type: necklace
Material: paper, paint, thread

砺行 / Honing

2021 北京国际
首饰艺术展
2021 BEIJING
INTERNATIONAL
JEWELLERY ART
EXHIBITION

194

团结
UNION

作者姓名: 孙月阳（中国）
作品类型: 项链
作品材质: 银

Artist: Sun Yueyang（China）
Type: necklace
Material: silver

锈花
BLOOMING RUST

作者姓名: 覃霄（中国）
作品类型: 其他
作品材质: 树脂，铁锈，铜锈

Artist: Qin Xiao（China）
Type: other
Material: resin, iron rust, copper rust

砺行 / Honing

2021 北京国际
首饰艺术展
2021 BEIJING
INTERNATIONAL
JEWELLERY ART
EXHIBITION

195

生——花生
LIFE — PEANUT

作者姓名: 唐超（中国）
作品类型: 戒指
作品材质: 银，黄铜

Artist: Tang Chao（China）
Type: ring
Material: silver, brass

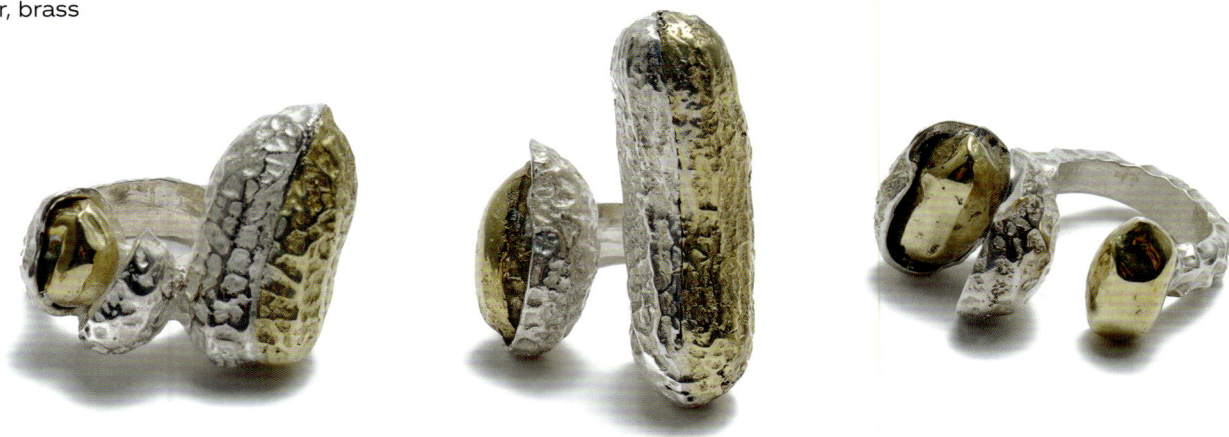

花语系列之"忆"
FLOWER LANGUAGE SERIES — MEMORY

作者姓名: 田伟玲（中国）
作品类型: 项链
作品材质: 银，布

Artist: Tian Weiling（China）
Type: necklace
Material: silver, cloth

砺行 / Honing

2021 北京国际
首饰艺术展
2021 BEIJING
INTERNATIONAL
JEWELLERY ART
EXHIBITION

196

离离
SPREADIG

作者姓名: 万紫薇（中国）
作品类型: 胸针
作品材质: 925 银，TPU 膜，皮布

Artist: Wan Ziwei（China）
Type: brooch
Material: 925 silver, TPU film, leather cloth

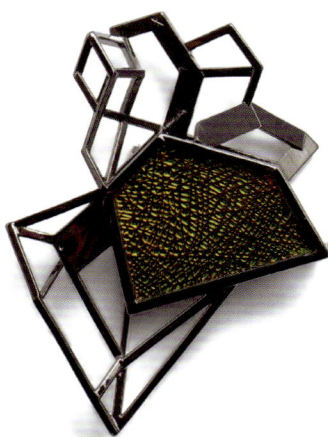

春华秋实
FRUITFUL ACHIEVEMENTS

作者姓名: 汪大江（中国）
作品类型: 胸针
作品材质: 金，925 银，和田玉

Artist: Wang Dajiang（China）
Type: brooch
Material: gold, 925 silver, nephrite

砺行 / Honing

2021 北京国际
首饰艺术展
2021 BEIJING
INTERNATIONAL
JEWELLERY ART
EXHIBITION

197

红色记忆
RED MEMORY

作者姓名: 汪正虹（中国）
作品类型: 项链
作品材质: 925 银，火山石，纪念章

Artist: Wang Zhenghong （China）
Type: necklace
Material: 925 silver, pelelith, medal

砺行 / Honing

2021 北京国际
首饰艺术展
2021 BEIJING
INTERNATIONAL
JEWELLERY ART
EXHIBITION

198

作者姓名： 王琲（中国）
作品类型： 头饰
作品材质： 陶瓷，925 银

Artist: Wang Bei（China）
Type: headwear
Material: ceramic, 925 silver

砺行 / Honing

2021 北京国际
首饰艺术展
2021 BEIJING
INTERNATIONAL
JEWELLERY ART
EXHIBITION

199

曲线
CURVE

作者姓名: 王浩睿（中国）
作品类型: 其他
作品材质: 银

Artist: Wang Haorui（China）
Type: other
Material: silver

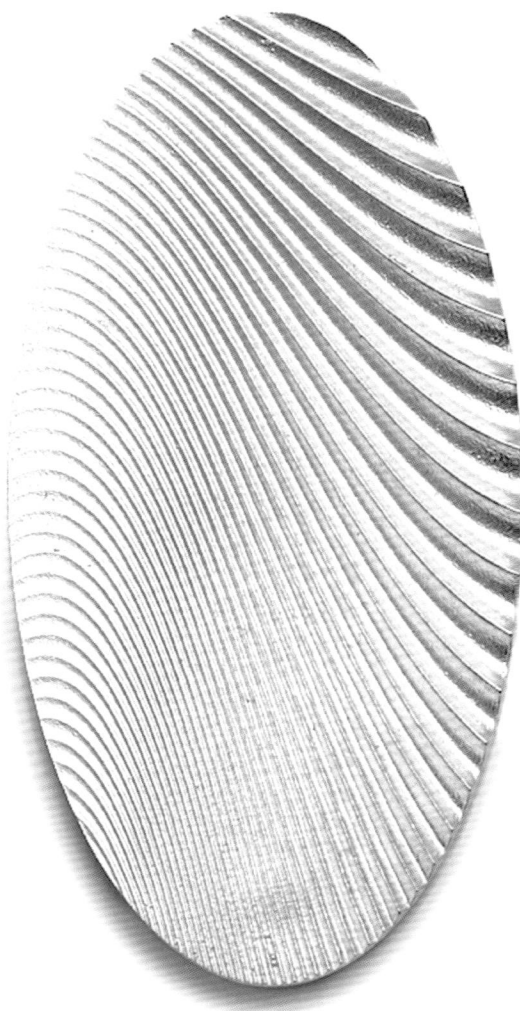

砺行 / Honing

2021 北京国际
首饰艺术展
2021 BEIJING
INTERNATIONAL
JEWELLERY ART
EXHIBITION

200

时间杀手
TIME KILLER

作者姓名:　王婧颖（中国）
作品类型:　戒指
作品材质:　银

Artist:　Wang Jingying（China）
Type:　ring
Material:　silver

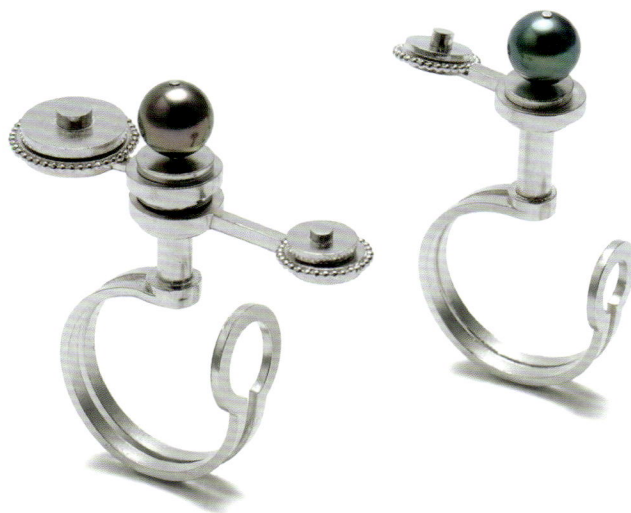

栖
HABITAT

作者姓名:　王敬（中国）
作品类型:　胸针
作品材质:　18K 黄金，银，航空铝，珍珠

Artist:　Wang Jing（China）
Type:　brooch
Material:　18K yellow gold, silver,
　　　　　aviation aluminum, pearl

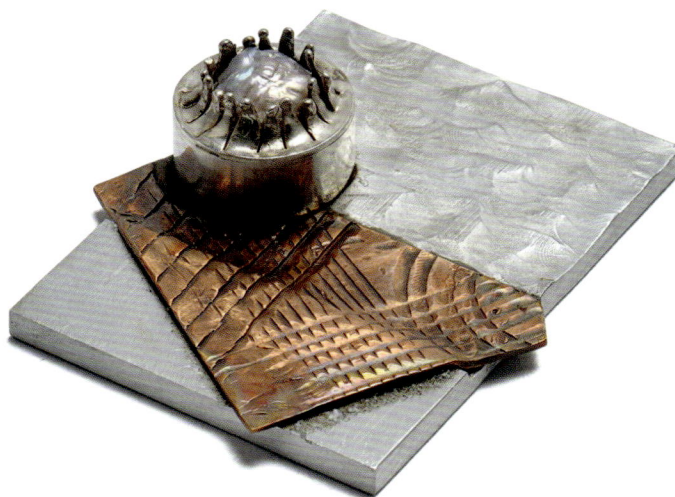

砺行 / Honing

2021 北京国际
首饰艺术展
2021 BEIJING
INTERNATIONAL
JEWELLERY ART
EXHIBITION

201

银行 1930
BANK 1930

作者姓名: **王靖（中国）**
作品类型: **胸针**
作品材质: **925 银，合成立方氧化锆**

Artist: **Wang Jing（China）**
Type: **brooch**
Material: **925 silver, synthetic cubic zirconia**

蝴蝶光斑
THE BUTTERFLY SPOT

作者姓名: **王连赛（中国）**
作品类型: **胸针**
作品材质: **钛金属，银镀 18K 黄金，玫瑰金，锆石**

Artist: **Wang Liansai（China）**
Type: **brooch**
Material: **titanium, 18K yellow gold plated silver, rose gold, zircon**

砺行 / Honing

2021 北京国际
首饰艺术展
2021 BEIJING
INTERNATIONAL
JEWELLERY ART
EXHIBITION

202

流光浮影
STREAMER FLOATING SHADOW

作者姓名:　王茜（中国）
作品类型:　胸针
作品材质:　珐琅，花丝，925 银

Artist:　　　Wang Qian（China）
Type:　　　brooch
Material:　　enamel, filaments, 925 silver

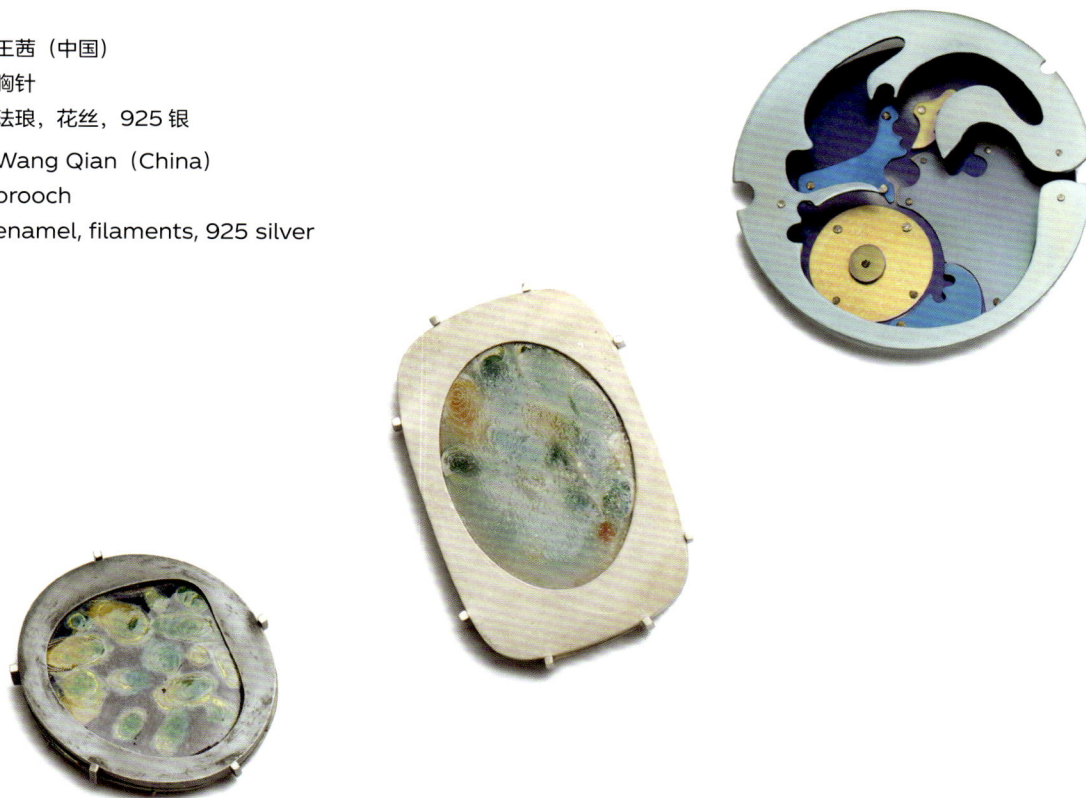

粉色的花 A
PINK FLOWER A

作者姓名:　王晴（中国）
作品类型:　胸针
作品材质:　黄铜，3D 打印透明树脂，颜料

Artist:　　　Wang Qing（China）
Type:　　　brooch
Material:　　brass, 3D printing transparent resin,
　　　　　　pigment

砺行 / Honing

2021 北京国际
首饰艺术展
2021 BEIJING
INTERNATIONAL
JEWELLERY ART
EXHIBITION

203

梦
DREAM

作者姓名: **王少军（中国）**
作品类型: **其他**
作品材质: **青白玉**

Artist: **Wang Shaojun（China）**
Type: **other**
Material: **white jade**

砺行 / Honing

2021 北京国际
首饰艺术展
2021 BEIJING
INTERNATIONAL
JEWELLERY ART
EXHIBITION

204

无题
UNTITLED

作者姓名: 王少军，刘建钊（中国）
作品类型: 其他
作品材质: 玉

Artist: Wang Shaojun, Liu Jianzhao（China）
Type: other
Material: jade

砺行 / Honing

2021 北京国际
首饰艺术展
2021 BEIJING
INTERNATIONAL
JEWELLERY ART
EXHIBITION

205

如梦之梦，金鱼，木槿
A DREAM LIKE A DREAM,
GOLDFISH, HIBISCUS FLOWERS

作者姓名: 王圣临（中国）

作品类型: 耳环，胸针

作品材质: 18K 黄金，坦桑石，钻石，蓝宝石，
鹦鹉羽毛，翡翠，红宝石，黄钻石，
水晶，蓝宝石，羽毛，尖晶石，孔雀羽毛

Artist: Wang Shenglin（China）

Type: earrings, brooch

Material: 18K yellow gold, tanzanite, diamond, sapphire,
parrot feather, emerald, ruby,
yellow diamond, crystal, feather,
spinel, peacock feather

砺行 / Honing

2021 北京国际
首饰艺术展
2021 BEIJING
INTERNATIONAL
JEWELLERY ART
EXHIBITION

206

融
MERGE

作者姓名: 王泰迪（中国）
作品类型: 胸针
作品材质: 欧洲古董首饰，银，老瓷片

Artist: Wang Taidi（China）
Type: brooch
Material: antique European jewellery, silver, old porcelain

砺行 / Honing

2021北京国际
首饰艺术展
2021 BEIJING
INTERNATIONAL
JEWELLERY ART
EXHIBITION

207

飘 · 舞 1
FLY · DANCE 1

作者姓名:	王文静（中国）
作品类型:	胸针
作品材质:	银镀 18K 金，彩色宝石
Artist:	Wang Wenjing（China）
Type:	brooch
Material:	18K gold plated silver, gemstones

砺行 / Honing

2021 北京国际
首饰艺术展
2021 BEIJING
INTERNATIONAL
JEWELLERY ART
EXHIBITION

208

灵魂砝码：千姿之重
SOUL WEIGHT: THE WEIGHT OF A THOUSAND POSES

作者姓名： 王晓昕（中国）
作品类型： 其他
作品材质： 不锈钢，紫铜，金

Artist: Wang Xiaoxin（China）
Type: other
Material: stainless steel, red copper, gold

砺行 / Honing

2021北京国际
首饰艺术展
2021 BEIJING
INTERNATIONAL
JEWELLERY ART
EXHIBITION

209

药嘉年华
PILL'S CARNIVAL

作者姓名: 王轩（中国）
作品类型: 项链
作品材质: 光敏树脂镀金属，环氧树脂

Artist: Wang Xuan（China）
Type: necklace
Material: photosensitive resin gold plating,
epoxy resin

砺行 / Honing

2021 北京国际
首饰艺术展
2021 BEIJING
INTERNATIONAL
JEWELLERY ART
EXHIBITION

210

奔赴——碎
TO — BROKEN

作者姓名: 王玄（中国）
作品类型: 胸针
作品材质: 铜黏土

Artist: Wang Xuan（China）
Type: brooch
Material: copper clay

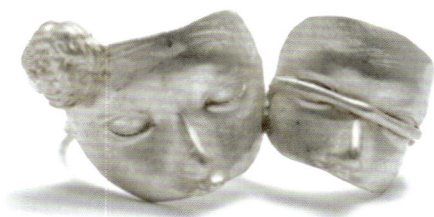

临行 / Honing

2021 北京国际
首饰艺术展
2021 BEIJING
INTERNATIONAL
JEWELLERY ART
EXHIBITION

211

蛹梦
PUPA DREAM

作者姓名: 王雪（中国）
作品类型: 手镯
作品材质: 银，银镀金

Artist: Wang Xue（China）
Type: bracelet
Material: silver, gold plated silver

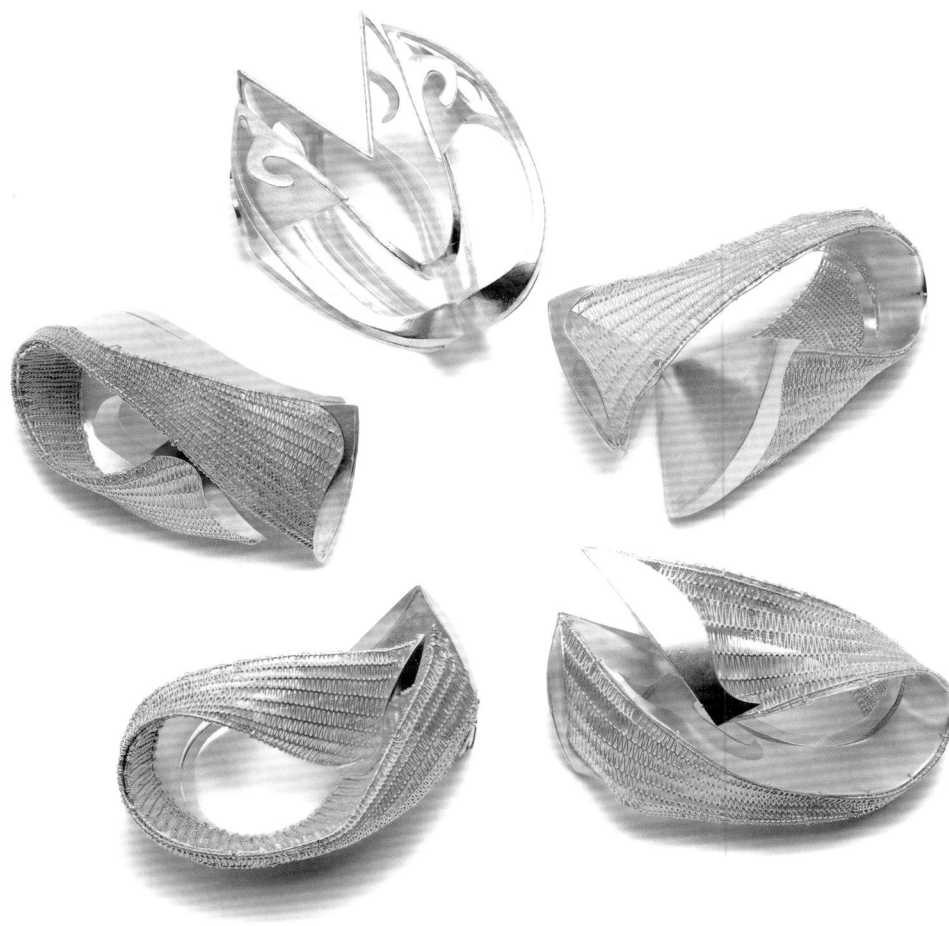

砺行 / Honing

2021 北京国际
首饰艺术展
2021 BEIJING
INTERNATIONAL
JEWELLERY ART
EXHIBITION

212

金玉其内
BEAUTIFUL INSIDE OF GOLD

作者姓名: 王亚萍（中国）
作品类型: 项链
作品材质: 浮石，925 银，合成立方氧化锆

Artist: Wang Yaping（China）
Type: necklace
Material: pumice, 925 silver,
synthetic cubic zirconia

砺行 / Honing

2021 北京国际
首饰艺术展
2021 BEIJING
INTERNATIONAL
JEWELLERY ART
EXHIBITION

213

无常（吊坠），无常（戒指）
IMPERMANENCE(PENDANT),
IMPERMANENCE(RING)

作者姓名： 王莹（中国）
作品类型： 项链，戒指
作品材质： 银

Artist: Wang Ying（China）
Type: necklace, ring
Material: silver

砺行 / Honing

2021 北京国际
首饰艺术展
2021 BEIJING
INTERNATIONAL
JEWELLERY ART
EXHIBITION

214

我的钻石在哪里 1，
我的钻石在哪里 2
WHERE ARE MY DIAMONDS 1,
WHERE ARE MY DIAMONDS 2

作者姓名： 王越（中国）
作品类型： 耳饰，其他
作品材质： 黄铜，925 银，锆石，羽毛

Artist: Wang Yue（China）
Type: earring, other
Material: brass, 925 silver,
zirconia, feather

融，窥
MELT, PEEP

作者姓名： 王哲昊（中国）
作品类型： 胸针，戒指
作品材质： 铝，玻璃，树脂，
软陶，网格布，银

Artist: Wang Zhehao（China）
Type: brooch, ring
Material: aluminum,
glass, resin, polymer clay,
interlaced fabric, silver

砺行 / Honing

2021 北京国际
首饰艺术展
2021 BEIJING
INTERNATIONAL
JEWELLERY ART
EXHIBITION

215

作者姓名: 吴芳（中国）
作品类型: 项链
作品材质: 银，树脂，钢丝

Artist: Wu Fang（China）
Type: necklace
Material: silver, resin, steel wire

砺行 / Honing

2021 北京国际
首饰艺术展
2021 BEIJING
INTERNATIONAL
JEWELLERY ART
EXHIBITION

216

翡翠与透明胶带，不抽烟，无珠宝
JADEITE AND SCOTCH TAPE, NO SMOKING, NO JEWELLERY

作者姓名： 魏子欣（中国）
作品类型： 项链，其他
作品材质： 翡翠，透明胶带，黄铜，红宝石，卷烟，锆石

Artist: Wei Zixin（China）
Type: necklace, other
Material: jadeite, scotch tape,
copper, ruby, cigarette, zircon

砺行 / Honing

2021 北京国际
首饰艺术展
2021 BEIJING
INTERNATIONAL
JEWELLERY ART
EXHIBITION

217

拉·李纳森特

LA RINASCENT

作者姓名: 温馨（中国）

作品类型: 胸针

作品材质: 18K 金，红宝石

Artist: Wen Xin（China）

Type: brooch

Material: 18K gold, ruby

乡恋

HOMESICK

作者姓名: 吴二强（中国）

作品类型: 胸针

作品材质: 925 银

Artist: Wu Erqiang（China）

Type: brooch

Material: 925 silver

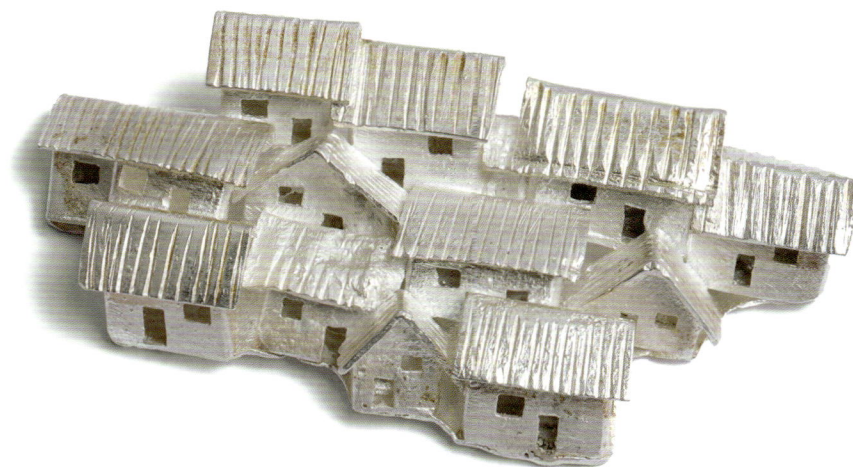

砺行 / Honing

2021 北京国际
首饰艺术展
2021 BEIJING
INTERNATIONAL
JEWELLERY ART
EXHIBITION

218

第零感
NO. ZERO SENSE

作者姓名：　**吴佳恒（中国）**
作品类型：　**其他**
作品材质：　**银**

Artist:　　**Wu Jiaheng（China）**
Type:　　　**other**
Material:　**silver**

硕行 / Honing

2021 北京国际
首饰艺术展
2021 BEIJING
INTERNATIONAL
JEWELLERY ART
EXHIBITION

219

一路风景
VIEW ALL THE WAY

作者姓名: **吴金金（中国）**
作品类型: **胸针**
作品材质: **银，珍珠，宝石**
Artist: **Wu Jinjin（China）**
Type: **brooch**
Material: **silver, pearl, gemstone**

砺行 / Honing

2021 北京国际
首饰艺术展
2021 BEIJING
INTERNATIONAL
JEWELLERY ART
EXHIBITION

220

壤1，壤2
EARTH 1, EARTH 2

作者姓名：　吴孟儒（中国）
作品类型：　胸针
作品材质：　自制合金，珐琅，不锈钢，金箔

Artist:　Wu Mengru（China）
Type:　brooch
Material:　homemade alloy, enamel,
　　　　　stainless steel, gold leaf

红色的希望，枯木逢春，初
RED HOPE, DEAD WOOD MEETS SPRING,
BEGINNING

作者姓名：　吴铭璐（中国）
作品类型：　胸针
作品材质：　纸浆，钉子，铜，棉线，
　　　　　纸，铁钉，金墨，树脂，银

Artist:　Wu Minglu（China）
Type:　brooch
Material:　pulp, nail, copper, cotton thread,
　　　　　paper, iron nail, gold ink, resin, silver

2021 北京国际
首饰艺术展
2021 BEIJING
INTERNATIONAL
JEWELLERY ART
EXHIBITION

221

莲连得子
LOTUS TO GET SON

作者姓名: 吴味霖（中国）
作品类型: 项链，胸针
作品材质: 银，锆石，鹿角

Artist: Wu Weilin（China）
Type: necklace, brooch
Material: silver, zircon, antler

冬蕴
WINTER

作者姓名: 吴越卓（中国）
作品类型: 胸针
作品材质: 钛金属，大漆，金箔，贝母

Artist: Wu Yuezhuo（China）
Type: brooch
Material: titanium, lacquer, gold leaf, fritillaria

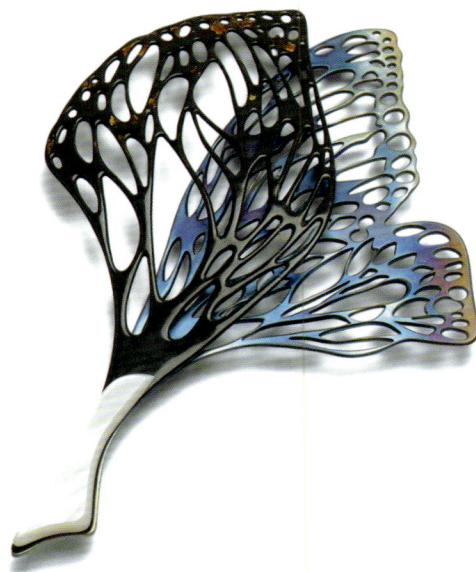

砺行 / Honing

2021 北京国际
首饰艺术展
2021 BEIJING
INTERNATIONAL
JEWELLERY ART
EXHIBITION

222

物欲横流 1，物欲横流 2，物欲横流 3
MATERIALISTIC 1, MATERIALISTIC 2,
MATERIALISTIC 3

作者姓名： 夏海原（中国）
作品类型： 手镯
作品材质： 925 银

Artist: Xia Haiyuan（China）
Type: bracelet
Material: 925 silver

蚌病生珠，"首饰"
MUSSEL BEADS, "JEWELLERY"

作者姓名： 肖传鑫（中国）
作品类型： 胸针
作品材质： 珍珠，钛，银，钢

Artist: Xiao Chuanxin（China）
Type: brooch
Material: pearl, titanium, silver, steel

砥行 / Honing

2021 北京国际
首饰艺术展
2021 BEIJING
INTERNATIONAL
JEWELLERY ART
EXHIBITION

223

华夏之歌
SONG OF CHINA

作者姓名: 肖尧（中国）
作品类型: 项链，耳饰
作品材质: 925 银，珍珠

Artist: Xiao Yao（China）
Type: necklace, earring
Material: 925 silver, pearl

器蕴
QI YUN

作者姓名: 谢媛（中国）
作品类型: 胸针
作品材质: 黄铜

Artist: Xie Yuan（China）
Type: brooch
Material: brass

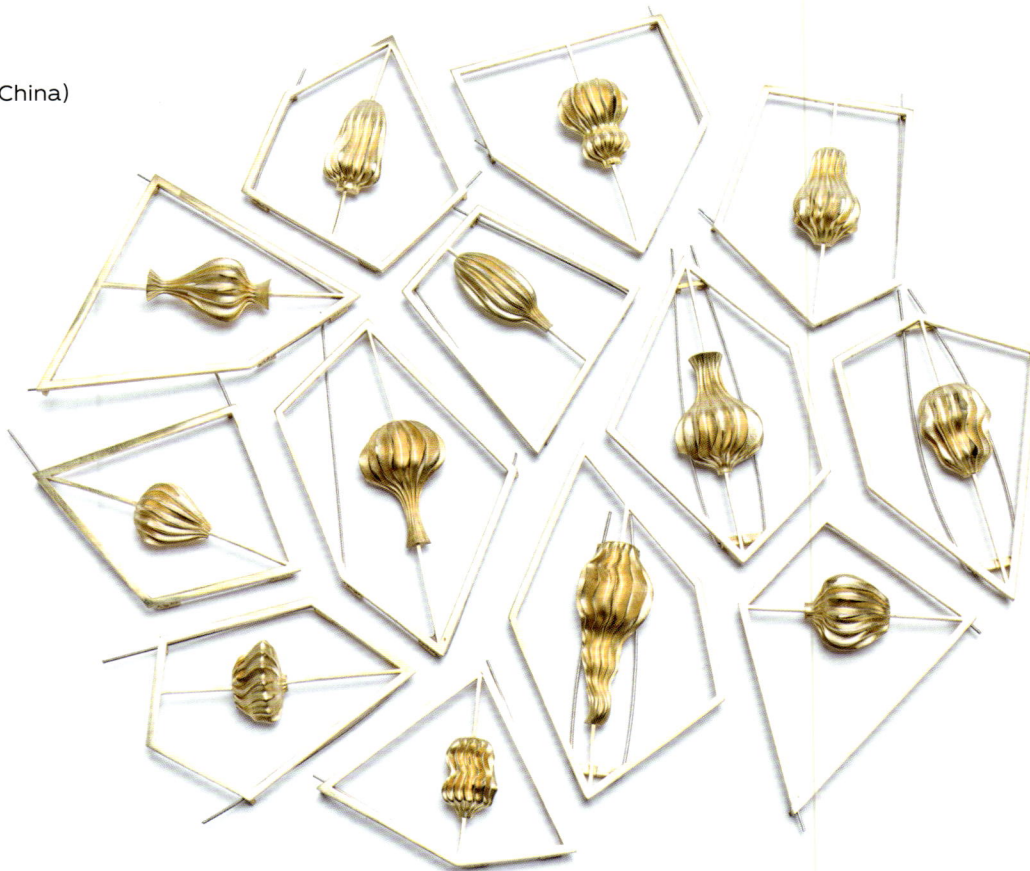

砺行 / Honing

2021 北京国际
首饰艺术展
2021 BEIJING
INTERNATIONAL
JEWELLERY ART
EXHIBITION

224

疏影
SHADOW OF ALIENATION

作者姓名: 谢赟怡（中国）
作品类型: 戒指
作品材质: 925 银镀金，锆石

Artist: Xie Yunyi（China）
Type: ring
Material: gold plated 925 silver, zircon

纪念碑
MONUMENTS

作者姓名: 徐程慧圆（中国）
作品类型: 胸针，项链
作品材质: 水泥，银，光导纤维，星光蓝宝石，丝绒布

Artist: Xu Chenghuiyuan（China）
Type: brooch, necklace
Material: cement, silver, optical fiber,
star sapphire, velvet cloth

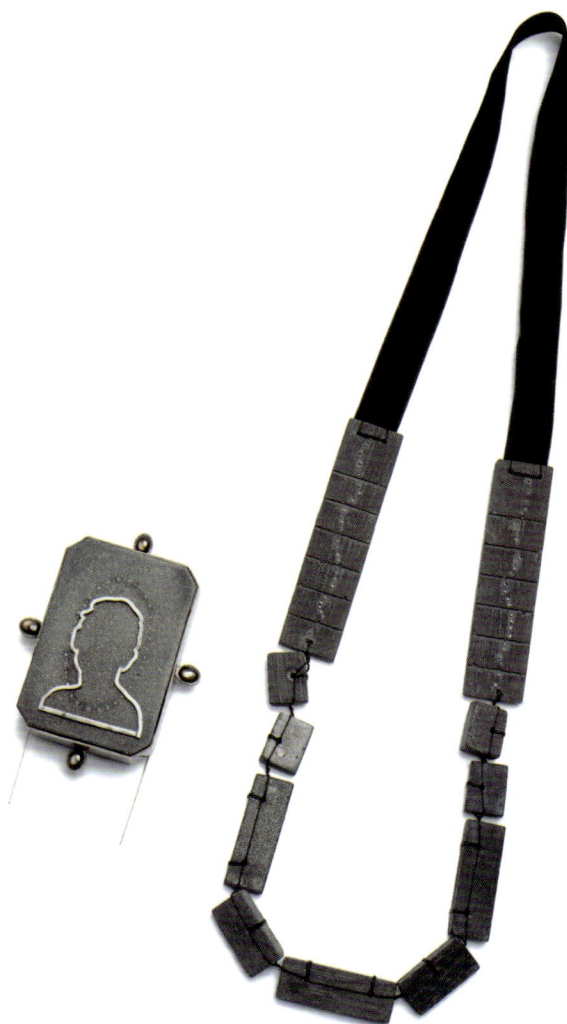

砺行 / Honing

2021 北京国际
首饰艺术展
2021 BEIJING
INTERNATIONAL
JEWELLERY ART
EXHIBITION

225

金戈铁马
HINING SPEARS & ARMORED HORSES

作者姓名: 熊芊芊（中国）
作品类型: 手镯
作品材质: 银，青玉

Artist: Xiong Dudu（China）
Type: bracelet
Material: silver, green jade

2021 北京国际
首饰艺术展
2021 BEIJING
INTERNATIONAL
JEWELLERY ART
EXHIBITION

指尖闪耀戒指 1，
指尖闪耀戒指 2，
指尖闪耀戒指 3
FINGERTIP SPARKLE RING 1,
FINGERTIP SPARKLE RING 2,
FINGERTIP SPARKLE RING 3

作者姓名： 徐可（中国）

作品类型： 戒指

作品材质： 白水晶，祖母绿，白钻，
18K 白金，彩色蓝宝，水晶片，
粉托帕石，绿碧玺，尖晶石

Artist: Xu Ke（China）

Type: ring

Material: white crystal, emerald, white diamond,
18K white gold, colored sapphire, crystal flake,
pink topaz, green tourmaline, spinel

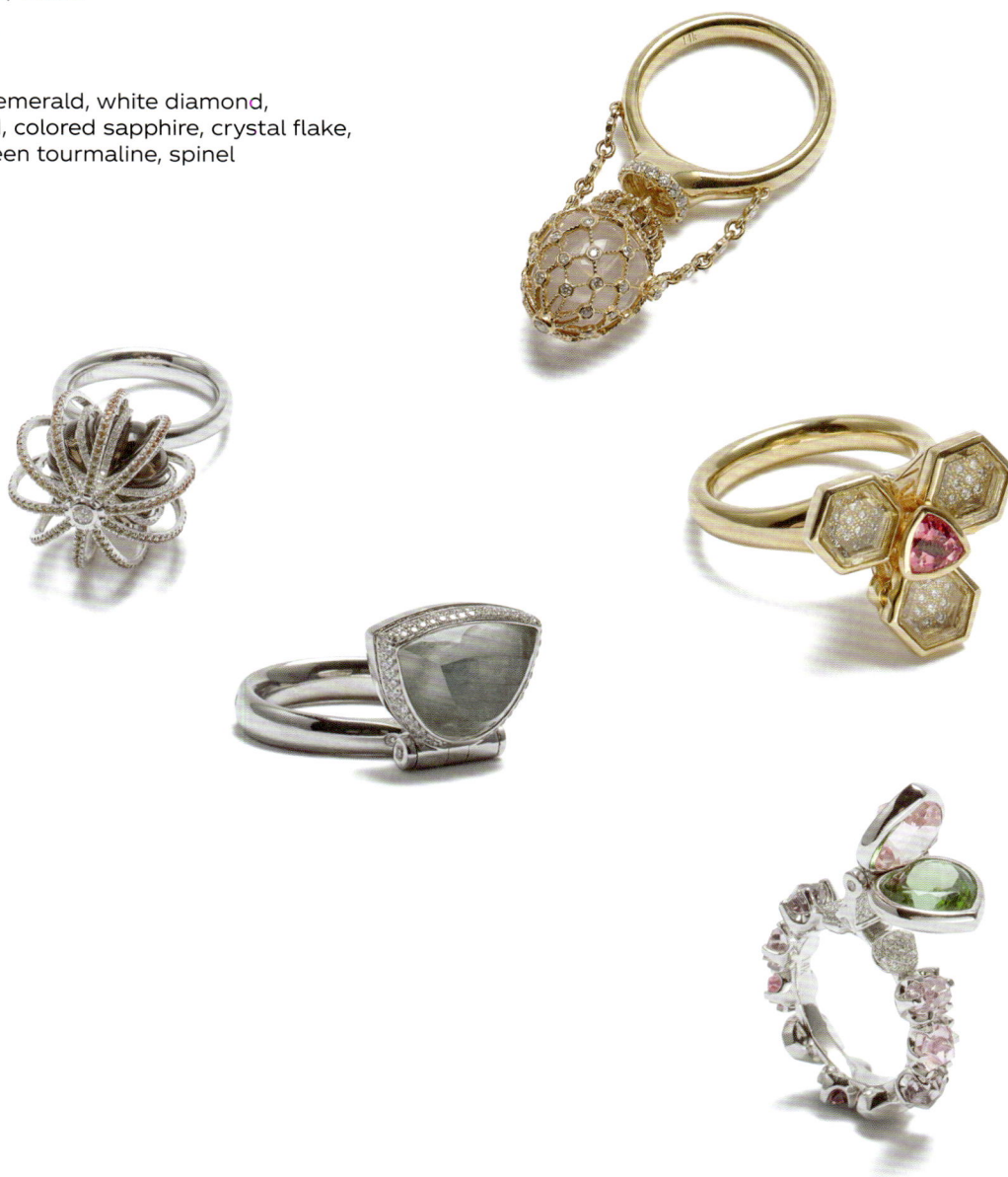

师行 / Honing

2021 北京国际
首饰艺术展
2021 BEIJING
INTERNATIONAL
JEWELLERY ART
EXHIBITION

227

走龙戏珠
WALKING DRAGON PLAY PEARL

作者姓名： 徐婧竞（中国）
作品类型： 项链
作品材质： 银，锆石，水晶

Artist: Xu Jingjing（China）
Type: necklace
Material: silver, zircon, crystal

承载胸针 1，
承载胸针 2，
承载胸针 3
BEARING BROOCH 1,
BEARING BROOCH 2,
BEARING BROOCH 3

作者姓名： 徐倩（中国）
作品类型： 胸针
作品材质： 银，纸，木板，光栅片，红宝石，鸡血石

Artist: Xu Qian（China）
Type: brooch
Material: silver, paper, wood panel, grating sheet, ruby, bloodstone

砺行 / Honing

2021 北京国际
首饰艺术展
2021 BEIJING
INTERNATIONAL
JEWELLERY ART
EXHIBITION

228

青山依旧1，青山依旧3
AOYAMA IS STILL 1, AND AOYAMA IS STILL 3

作者姓名： 徐翔宇（中国）
作品类型： 项链，手镯
作品材质： 煤渣，黄土，铝

Artist: Xu Xiangyu（China）
Type: necklace, bracelet
Material: cinder, loess, aluminum

俄罗斯方块系列
"TETRIS" SERIES

作者姓名： 许国蕤（中国）
作品类型： 胸针，耳饰
作品材质： 925 银电镀黑金，立方氧化锆

Artist: Xu Guorui（China）
Type: brooch, earring
Material: black gold plated 925 silver,
cubic zirconia

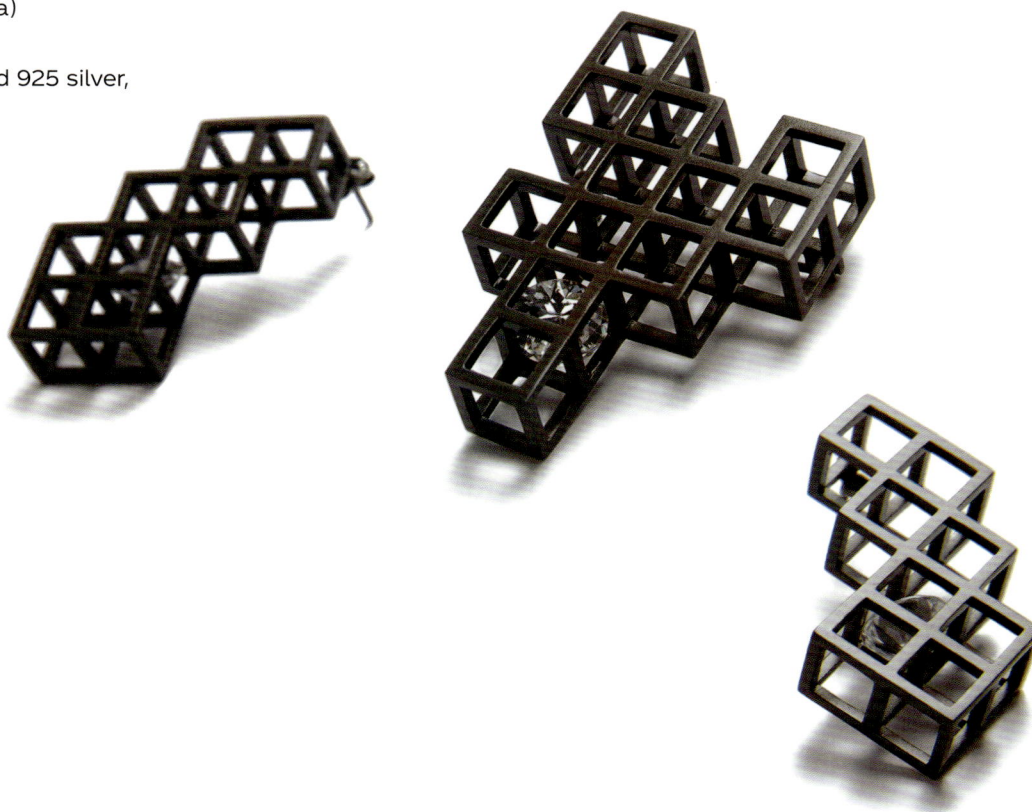

砺行 / Honing

2021 北京国际
首饰艺术展
2021 BEIJING
INTERNATIONAL
JEWELLERY ART
EXHIBITION

229

文明
CIVILIZATION

作者姓名:　许延平（中国）
作品类型:　项链
作品材质:　南红，银，砂，金

Artist:　　Xu Yanping（China）
Type:　　　necklace
Material:　nanjiang carnelian,
　　　　　silver, sand, gold

曲线混接
BLEND CURVE

作者姓名:　薛菲　（中国）
作品类型:　手镯，胸针
作品材质:　玉石，黄铜

Artist:　　Xue Fei　（China）
Type:　　　bracelet, brooch
Material:　jade, brass

砺行 / Honing

2021 北京国际
首饰艺术展
2021 BEIJING
INTERNATIONAL
JEWELLERY ART
EXHIBITION

230

尾系列，万物有生·浪舞，尾·镯
"TAIL" SERIES, ALL THINGS HAVE LIFE & WAVE DANCE, TAIL BRACELET

作者姓名： 薛雅昕（中国）

作品类型： 戒指，胸针，手镯

作品材质： 银，银镀金，蓝宝石，珍珠，
晶体颜料，石榴石，黄铜

Artist: Xue Yaxin（China）

Type: ring, brooch, bracelet

Material: silver, gold plated silver, sapphire,
pearl, crystal pigment, garnet, brass

繁衍
MULTIPLY

作者姓名： 闫黎（中国）

作品类型： 项链，耳饰

作品材质： 银镀 18K 白金，钻石，珍珠

Artist: Yan Li（China）

Type: necklace, earring

Material: 18K white gold plated silver,
diamond, pearl

砺行 / Honing

2021 北京国际
首饰艺术展
2021 BEIJING
INTERNATIONAL
JEWELLERY ART
EXHIBITION

231

"织" 然而然
WEAVING IS NATURAL

作者姓名: 闫政旭（中国）
作品类型: 胸针，项链
作品材质: 纯银，925 银

Artist: Yan Zhengxu（China）
Type: brooch, necklace
Material: sterling silver, 925 silver

十二庭院
TWELVE YARDS

作者姓名: 严政秋（中国）
作品类型: 胸针
作品材质: 紫铜，黄铜，银

Artist: Yan Zhengqiu（China）
Type: brooch
Material: red copper, brass, silver

砺行 / Honing

2021 北京国际
首饰艺术展
2021 BEIJING
INTERNATIONAL
JEWELLERY ART
EXHIBITION

232

彼 · 路
THE OTHER ROAD

作者姓名： 颜聪聪（中国）
作品类型： 项链，胸针
作品材质： 天然大漆，纸，布，金粉

Artist: Yan Congcong（China）
Type: necklace, brooch
Material: natural lacquer, paper,
cloth, gold powder

砺行 / Honing

2021 北京国际
首饰艺术展
2021 BEIJING
INTERNATIONAL
JEWELLERY ART
EXHIBITION

233

晓花 1，晓花 2
DAWNING BLOSSOM 1,
DAWNING BLOSSOM 2

作者姓名： 杨漫（中国）
作品类型： 项链
作品材质： 925 银，翡翠

Artist: Yang Man（China）
Type: necklace
Material: 925 silver, jadeite

砺行 / Honing

2021 北京国际
首饰艺术展
2021 BEIJING
INTERNATIONAL
JEWELLERY ART
EXHIBITION

234

流光溢彩
AMBILIGHT

作者姓名： 杨烨（中国）
作品类型： 胸针
作品材质： 钛金属
Artist: Yang Ye（China）
Type: brooch
Material: titanium

生于自然 1，生于自然 2，生于自然 3
BORN OF NATURE 1，BORN OF NATURE 2，
BORN OF NATURE 3

作者姓名： 杨逸伦（中国）
作品类型： 胸针
作品材质： 贝母，钛，珍珠，不锈钢
Artist: Yang Yilun（China）
Type: brooch
Material: fritillaria, titanium, pearl, stainless steel

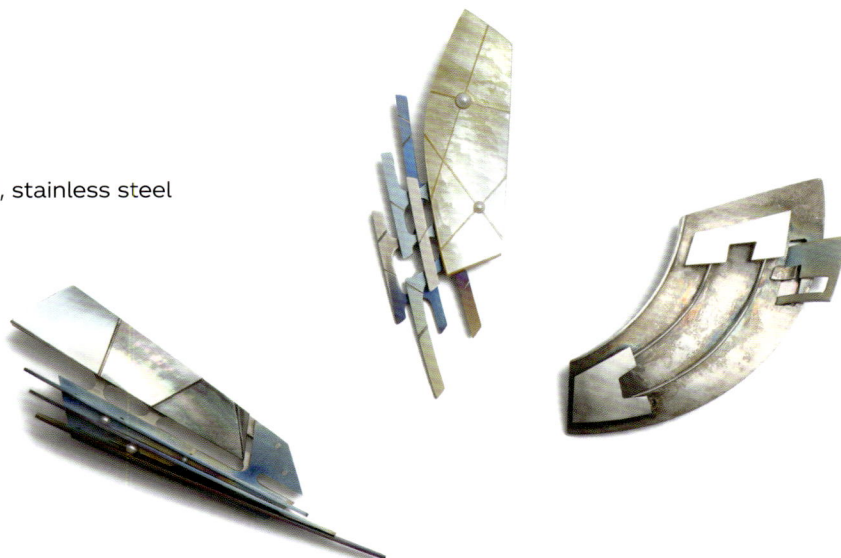

砺行 / Honing

2021 北京国际
首饰艺术展
2021 BEIJING
INTERNATIONAL
JEWELLERY ART
EXHIBITION

235

量材度理，定心而行，规矩方圆
MEASURE THE MATERIAL, CENTER, AND BE REGULAR

作者姓名： 杨钊（中国）
作品类型： 胸针
作品材质： 银镀金，宝石

Artist: Yang Zhao（China）
Type: brooch
Material: gold plated silver, gemstone

补钙
CALCIUM SUPPLEMENTATION

作者姓名： 杨中雄（中国）
作品类型： 项链
作品材质： 银，猪骨，狗链

Artist: Yang Zhongxiong（China）
Type: necklace
Material: silver, swine bone, dog chain

砥行 / Honing

2021北京国际
首饰艺术展
2021 BEIJING
INTERNATIONAL
JEWELLERY ART
EXHIBITION

236

蓝
BLUE

作者姓名： 姚西莹（中国）
作品类型： 项链
作品材质： 钛，黄铜，银，锆石，
亚克力，不锈钢，布料

Artist: Yao Xiying（China）
Type: necklace
Material: titanium, brass, silver, zircon,
acrylic, stainless steel, cloth

护身符—— 白菜价，护身符—— 出入平安
AMULET — LOW PRICE,
AMULET — IN AND OUT SAFELY

作者姓名： 余晨玮（中国）
作品类型： 项链，其他
作品材质： 包装盒，棉片，口罩绳

Artist: Yu Chenwei（China）
Type: necklace, other
Material: packing box, cotton pad, mask cord

砺行 / Honing

2021 北京国际
首饰艺术展
2021 BEIJING
INTERNATIONAL
JEWELLERY ART
EXHIBITION

237

窥
PEEP

作者姓名: 袁梦（中国）
作品类型: 胸针
作品材质: 不锈钢，925 银，金箔

Artist: Yuan Meng（China）
Type: brooch
Material: stainless steel, 925 silver, gold foil

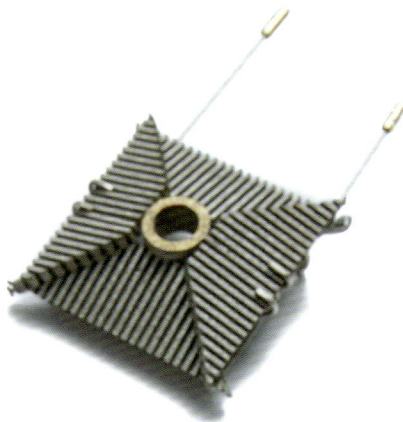

合
TOGETHER

作者姓名: 袁塔拉（中国）
作品类型: 项链
作品材质: 925 银，银镀 18K 金，翡翠，综合材料

Artist: Yuan Tala（China）
Type: necklace
Material: 925 silver, 18K gold plated silver,
jadeite, synthetic material

砥行 / Honing

2021 北京国际
首饰艺术展
2021 BEIJING
INTERNATIONAL
JEWELLERY ART
EXHIBITION

238

生命的印记 1，生命的印记 2，生命的印记 3
MARK OF LIFE 1, MARK OF LIFE 2, MARK OF LIFE 3

作者姓名：	袁文娟（中国）
作品类型：	项链
作品材质：	铝着色，树脂，925 银
Artist:	Yuan Wenjuan（China）
Type:	necklace
Material:	oxidized aluminum, resin, 925 silver

共生
SYMBIOSIS

作者姓名：	袁小伟（中国）
作品类型：	戒指
作品材质：	925 银，松木，水性木器漆
Artist:	Yuan Xiaowei（China）
Type:	ring
Material:	925 silver, pine, water-based wood lacquer

砺行 / Honing

2021 北京国际
首饰艺术展
2021 BEIJING
INTERNATIONAL
JEWELLERY ART
EXHIBITION

239

心枷锁——几何耳饰
HEART CHAINS — GEOMETRIC EARRINGS

作者姓名： 袁叶子（中国）
作品类型： 耳饰
作品材质： 银镀金

Artist: Yuan Yezi（China）
Type: earring
Material: gold plated silver

灯烛萤煌，灯火辉煌，灯影幢幢
LIGHTS AND CANDLES ARE BRILLIANT,
LIGHTS ARE BRILLIANT,
LIGHTS AND SHADOWS LOOK LIKE BUILDINGS

作者姓名： 袁祎彬（中国）
作品类型： 项链，耳饰，戒指
作品材质： 黄铜，玻璃，树脂

Artist: Yuan Yibin（China）
Type: necklace, earring, ring
Material: brass, glass, resin

砺行 / Honing

2021 北京国际
首饰艺术展
2021 BEIJING
INTERNATIONAL
JEWELLERY ART
EXHIBITION

240

旋玉蟠龙
SPIRAL JADE PAN DRAGON

作者姓名：　张凡（中国）
作品类型：　项链
作品材质：　和田玉，紫铜鎏金银

Artist:　　Zhang Fan（China）
Type:　　　necklace
Material:　 nephrite,
　　　　　　red copper glided
　　　　　　with gold and silver

彝韵·三角梅大号胸针，
彝韵·三角梅小号胸针，
彝韵·三角梅开口戒指
YI YUN TRIANGLE PLUM LARGE BROOCH,
YI YUN TRIANGLE PLUM SMALL BROOCH,
YI YUN TRIANGLE PLUM OPEN RING

作者姓名：　张继琳（中国）
作品类型：　胸针，戒指
作品材质：　铜镀 18K 白金，三醋酸面料

Artist:　　Zhang Jilin（China）
Type:　　　brooch, ring
Material:　 18K white gold plated copper,
　　　　　　triacetate fabric

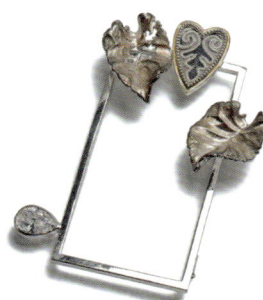

砺行 / Honing

2021 北京国际
首饰艺术展
2021 BEIJING
INTERNATIONAL
JEWELLERY ART
EXHIBITION

241

思愁之胸针，
思愁之戒指，
思愁之耳饰

THINK OF SORROW BROOCH,
THINK OF SORROW RING,
THINK OF SORROW EARRINGS

作者姓名： 张洁（中国）
作品类型： 胸针，戒指，耳饰
作品材质： 银，大漆，铜丝网，珍珠

Artist: Zhang Jie（China）
Type: brooch, ring, earring
Material: silver, lacquer, copper wire mesh, pearl

伊卡洛斯的邂逅，
伊卡洛斯的骨如意，
伊卡洛斯的翡翠烛屏

THE ENCOUNTER OF ICARUS,
THE BONE OF ICARUS,
THE EMERALD CANDLE SCREEN OF ICARUS

作者姓名： 张介安（中国）
作品类型： 项链，其他
作品材质： 铜，18K 黄金，银，皮革材料，
　　　　　 骨制品，兽爪，兽骨，翡翠

Artist: Zhang Jiean（China）
Type: necklace, other
Material: copper, 18K yellow gold, silver,
leather material, bone product,
animal claw, animal bone, jadeite

临行 / Honing

2021 北京国际
首饰艺术展
2021 BEIJING
INTERNATIONAL
JEWELLERY ART
EXHIBITION

242

花篓
FLOWER BASKET

作者姓名： 张磊（中国）
作品类型： 项链
作品材质： 925 银，大米，珍珠，丝绸

Artist: Zhang Lei（China）
Type: necklace
Material: 925 silver, rice, pearl, silk

一半，维吉娜，像素游戏
HALF, VIRGINA, PIXEL GAME

作者姓名： 张楠楠（中国）
作品类型： 胸针
作品材质： 银，银镀 18K 黄金，锆石，铜做旧，
黄色锆石，亚克力，不锈钢螺钉螺母

Artist: Zhang Nannan（China）
Type: brooch
Material: silver, 18K yellow gold plated silver, zircon,
distressed copper, yellow zircon, acrylic,
stainless steel screw and nut

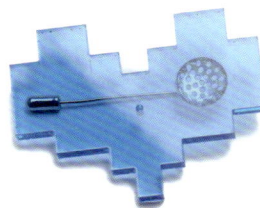

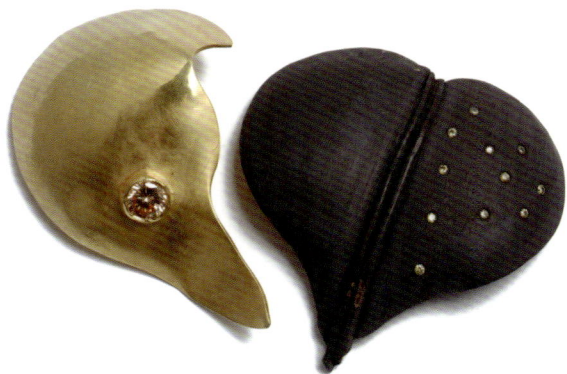

砺行 / Honing

2021 北京国际
首饰艺术展
2021 BEIJING
INTERNATIONAL
JEWELLERY ART
EXHIBITION

243

遥远的居民
DISTANT INHABITANTS

作者姓名:	张勤（中国）
作品类型:	胸针，耳饰
作品材质:	纯银丝和合金丝钩编，纯银，不锈钢丝，合金丝钩编，氧化银

Artist:	Zhang Qin（China）
Type:	brooch, earring
Material:	sterling silver and alloy wires crochet, sterling silver, stainless steel wire, alloy wire crochet, silver oxide

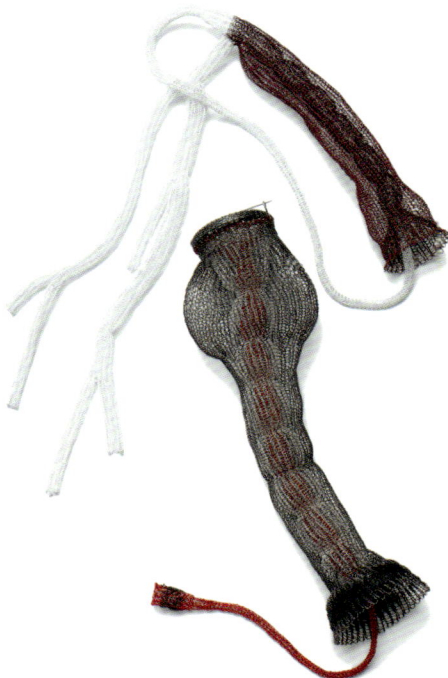

山海情，合鸣，契
MOUNTAINS AND SEAS, HARMONY, QI

作者姓名:	张斯婧（中国）
作品类型:	胸针
作品材质:	和田玉，18K 黄金

Artist:	Zhang Sijing（China）
Type:	brooch
Material:	nephrite, 18K yellow gold

砺行 / Honing

2021 北京国际
首饰艺术展
2021 BEIJING
INTERNATIONAL
JEWELLERY ART
EXHIBITION

244

秋·绪
AUTUMN THOUGHT

作者姓名:　张荣红（中国）
作品类型:　胸针
作品材质:　银

Artist:　　Zhang Ronghong（China）
Type:　　　brooch
Material:　silver

砺行 / Honing

2021 北京国际
首饰艺术展
2021 BEIJING
INTERNATIONAL
JEWELLERY ART
EXHIBITION

245

春 · 逸，岚，可餐
SPRING JOY, THE MIST IN THE MOUNTAINS, EATABLE

作者姓名： 张天亮（中国）
作品类型： 项链，耳饰
作品材质： 925 银，树脂，丝绸，钛金属，黄铜，黑绳

Artist: Zhang Tianliang（China）
Type: necklace, earring
Material: 925 silver, resin, silk,
titanium, brass, black rope

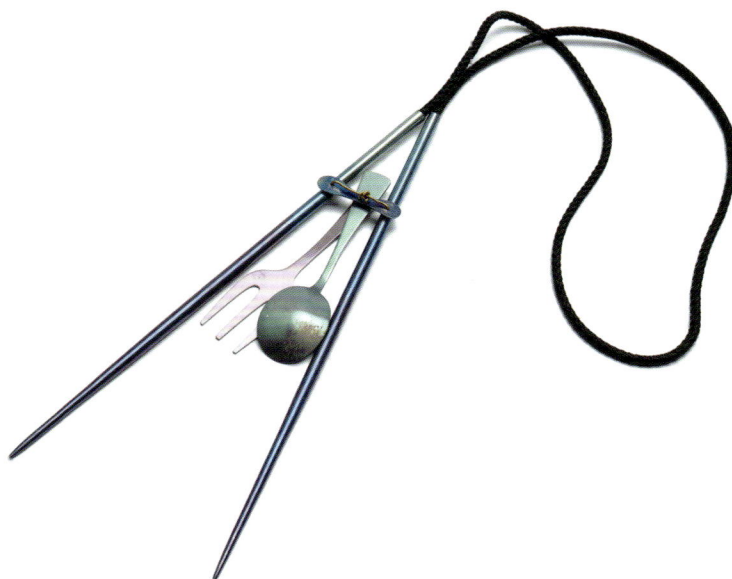

婺气
WU QI

作者姓名： 张文龙（中国）
作品类型： 项链，头饰
作品材质： 银

Artist: Zhang Wenlong（China）
Type: necklace, headdress
Material: silver

砺行 / Honing

2021 北京国际
首饰艺术展
2021 BEIJING
INTERNATIONAL
JEWELLERY ART
EXHIBITION

246

海洋精神第三部分
SPIRIT OF THE SEA, PART III

作者姓名: 张雯迪（中国）
作品类型: 胸针，项链
作品材质: 银，珐琅，纤维

Artist: Zhang Wendi（China）
Type: brooch, necklace
Material: silver, enamel, fiber

安菲尔
AMPHI

作者姓名: 张雪梅（中国）
作品类型: 胸针
作品材质: 纸

Artist: Zhang Xuemei（China）
Type: brooch
Material: paper

砺行 / Honing

2021 北京国际
首饰艺术展
2021 BEIJING
INTERNATIONAL
JEWELLERY ART
EXHIBITION

247

钻石恒久远，一颗永流传
A DIAMOND IS FOREVER

作者姓名： 张祎桐（中国）
作品类型： 其他
作品材质： 18K 白金，钻石，戒指盒，典当行包装袋

Artist: Zhang Yitong（China）
Type: other
Material: 18K white gold, diamond,
ring box, pawnshop packaging

竹劲 1，竹劲 2，竹劲 3
ZHUJING 1, ZHUJING 2, ZHUJING 3

作者姓名： 张译丹（中国）
作品类型： 胸针，耳饰
作品材质： 银，黄铜，玉

Artist: Zhang Yidan（China）
Type: brooch, earring
Material: silver, brass, jade

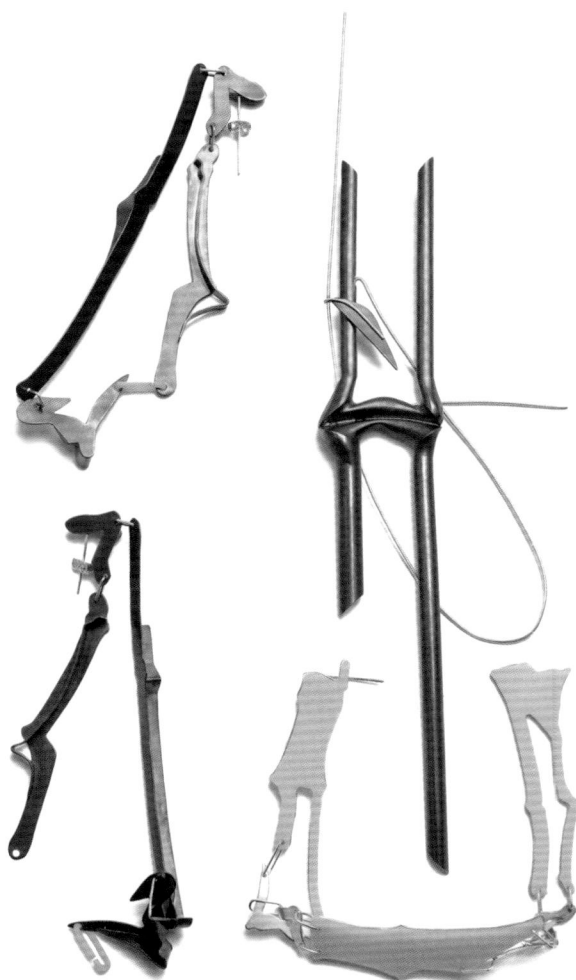

砺行 / Honing

2021 北京国际
首饰艺术展
2021 BEIJING
INTERNATIONAL
JEWELLERY ART
EXHIBITION

248

回声 1，回声 2，回声 3
ECHO 1，ECHO 2，ECHO 3

作者姓名：　张逸轩（中国）
作品类型：　手镯，耳饰，其他
作品材质：　银

Artist:　　　Zhang Yixuan（China）
Type:　　　 bracelet, earring, other
Material:　　silver

梦中的童话
FAIRY TALES IN THE DREAM

作者姓名：　章雨晴（中国）
作品类型：　胸针
作品材质：　紫铜，珐琅，纯银，不锈钢针
Artist:　　　Zhang Yuqing（China）
Type:　　　 brooch
Material:　　copper, enamel, sterling silver,
　　　　　　 stainless steel needle

砺行 / Honing

2021 北京国际
首饰艺术展
2021 BEIJING
INTERNATIONAL
JEWELLERY ART
EXHIBITION

249

伤痕首饰
SCARS JEWELLERY

作者姓名： 赵际婷（中国）
作品类型： 其他
作品材质： 银，珐琅等

Artist: Zhao Jiting（China）
Type: other
Material: silver, enamel, etc.

叠加，碎·整，视线
SUPERIMPOSED, BROKEN AND TIDY, LINE OF SIGHT

作者姓名： 赵建成（中国）
作品类型： 胸针，其他
作品材质： 鸡蛋壳，透明树脂

Artist: Zhao Jiancheng（China）
Type: brooch, other
Material: eggshell, transparent resin

砺行 / Honing

2021 北京国际
首饰艺术展
2021 BEIJING
INTERNATIONAL
JEWELLERY ART
EXHIBITION

250

破碎的时间，七种孤独，我们的天空
BROKEN TIME,
SEVEN KINDS OF LONELINESS, OUR SKY

作者姓名： 赵剑侠（中国）
作品类型： 胸针，其他
作品材质： 925 银，亚克力，竹，银

Artist: Zhao Jianxia（China）
Type: brooch, other
Material: 925 silver, acrylic, bamboo, silver

珍奇挚爱
RARE LOVE

作者姓名： 赵世笺（中国）
作品类型： 手镯，耳饰
作品材质： 925 银，Akoya 海水珍珠

Artist: Zhao Shijian（China）
Type: bracelet, earring
Material: 925 silver, Akoya sea pearl

砺行 / Honing

2021 北京国际
首饰艺术展
2021 BEIJING
INTERNATIONAL
JEWELLERY ART
EXHIBITION

251

大海在那头
THE SEA IS OVER THERE

作者姓名： 赵祎（中国）
作品类型： 胸针
作品材质： 木头，大漆，银，真丝，螺钿

Artist: Zhao Yi（China）
Type: brooch
Material: wood, lacquer, silver, silk, inlay

吾行 / Honing

2021 北京国际
首饰艺术展
2021 BEIJING
INTERNATIONAL
JEWELLERY ART
EXHIBITION

252

作者姓名： 郑静（中国）
作品类型： 项链
作品材质： 纸，铜，银

Artist: Zheng Jing（China）
Type: necklace
Material: paper, copper, silver

砺行 / Honing

2021 北京国际
首饰艺术展
2021 BEIJING
INTERNATIONAL
JEWELLERY ART
EXHIBITION

253

恐慌感与白系列
TREPIDATION AND WHITE SERIES

作者姓名： 赵莹（中国）
作品类型： 胸针
作品材质： 银，珐琅，木头，金箔，铜

Artist: Zhao Ying（China）
Type: brooch
Material: silver, enamel, wood,
gold leaf, copper

对戒，融化的爱
RINGS, MELTED LOVE

作者姓名： 郑婷玉（中国）
作品类型： 戒指，其他
作品材质： 银

Artist: Zheng Tingyu（China）
Type: ring, other
Material: silver

砺行 / Honing

2021 北京国际
首饰艺术展
2021 BEIJING
INTERNATIONAL
JEWELLERY ART
EXHIBITION

254

守·护，双笙
GUARD AND PROTECT, SHUANG SHENG

作者姓名： 郑稀文（中国）

作品类型： 胸针，耳饰

作品材质： 18K 金，大溪地黑珍珠，南洋浓金珍珠，
Akoya 珍珠，白钻，黑钻，祖母绿，沙弗莱

Artist: Zheng Xiwen（China）

Type: brooch, earring

Material: 18K gold, tahitian black pearl,
south sea concentrated golden pearl,
Akoya pearl, white diamond,
black diamond, emerald, tsavorite

念项链，念胸针，念耳环
"NIAN" NECKLACE, "NIAN" BROOCH, "NIAN" EARRING

作者姓名： 郑妍芳（中国）

作品类型： 项链，胸针，耳饰

作品材质： 银，竹，气球

Artist: Zheng Yanfang（China）

Type: necklace, brooch, earring

Material: silver, bamboo, balloon

听行 / Honing

2021 北京国际
首饰艺术展
2021 BEIJING
INTERNATIONAL
JEWELLERY ART
EXHIBITION

255

开在我心中的一朵苏打泡沫花儿
SODA — A FLOWER ROOTS IN MY HEART

作者姓名: 郑媛（中国）
作品类型: 其他
作品材质: 925 银，锆石

Artist: Zheng Yuan（China）
Type: other
Material: 925 silver, zircon

生生之地 · 禾 1,
生生之地 · 禾 2,
生生之地 · 禾 3
THE LAND OF LIFE — HE 1,
THE LAND OF LIFE — HE 2,
THE LAND OF LIFE — HE 3

作者姓名: 周若雪（中国）
作品类型: 项链，胸针
作品材质: 925 银

Artist: Zhou Ruoxue（China）
Type: necklace, brooch
Material: 925 silver

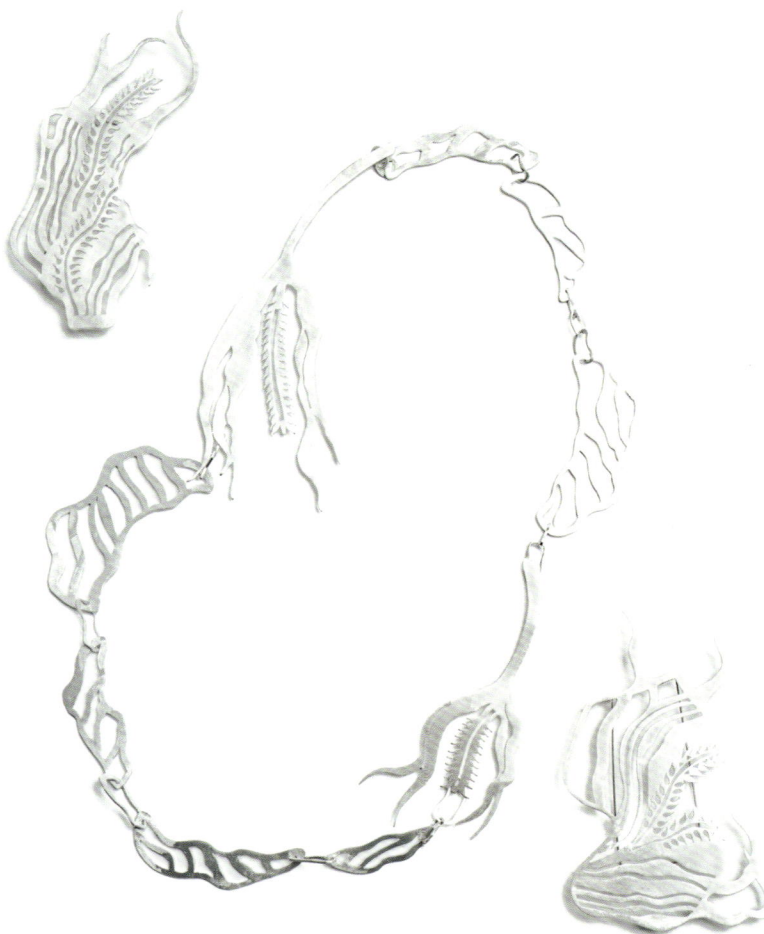

砺行 / Honing

2021 北京国际
首饰艺术展
2021 BEIJING
INTERNATIONAL
JEWELLERY ART
EXHIBITION

256

花雨
FLOWER RAIN

作者姓名： 周思成（中国）
作品类型： 胸针
作品材质： 银，蜜蜡，和田玉

Artist: Zhou Sicheng（China）
Type: brooch
Material: silver, beeswax, nephrite

遇见神鹿
MEET THE DEER GOD

作者姓名： 周思嘉（中国）
作品类型： 项链
作品材质： 银镀金，彩色合成立方氧化锆，珐琅

Artist: Zhou Sijia（China）
Type: necklace
Material: gold plated silver,
color synthetic cubic zirconia, enamel

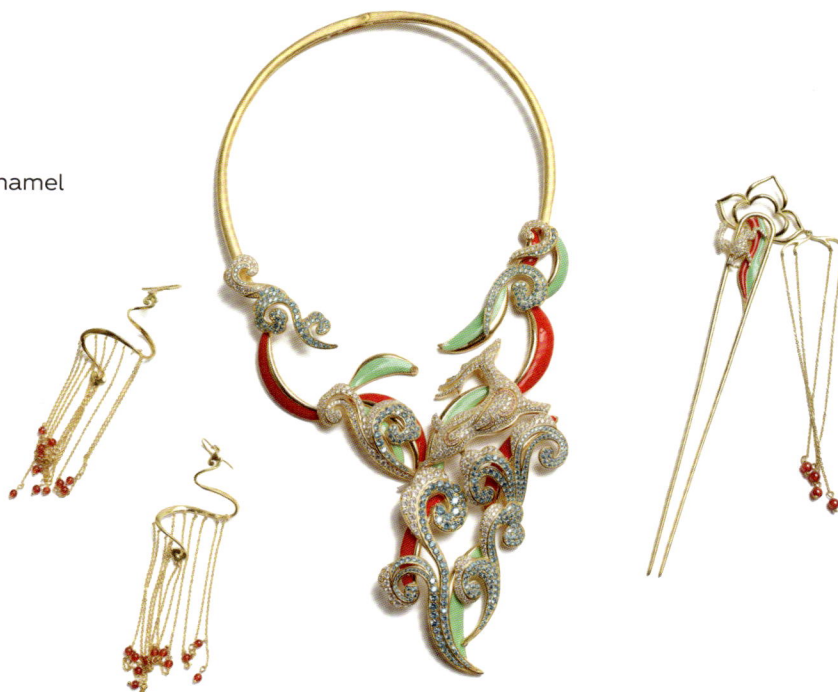

砺行 / Honing

2021 北京国际
首饰艺术展
2021 BEIJING
INTERNATIONAL
JEWELLERY ART
EXHIBITION

257

圆满系列 2，圆满系列 3
"SATISFACTORY" COLLECTION 2 ,
"SATISFACTORY" COLLECTION 3

作者姓名:	周潇（中国）
作品类型:	其他
作品材质:	银，尼龙
Artist:	Zhou Xiao（China）
Type:	other
Material:	silver, nylon

被束缚的自然 1，
被束缚的自然 2，
被束缚的自然 3
BOUND NATURE 1,
BOUND NATURE 2,
BOUND NATURE 3

作者姓名:	周震（中国）
作品类型:	胸针
作品材质:	银，湖石，不锈钢
Artist:	Zhou Zhen（China）
Type:	brooch
Material:	silver, lake stone, stainless steel

砥行 / Honing

2021 北京国际
首饰艺术展
2021 BEIJING
INTERNATIONAL
JEWELLERY ART
EXHIBITION

258

造字，奏
MAKING WORDS, PLAYING

作者姓名： 周子琪（中国）
作品类型： 其他
作品材质： 银，零件，纸浆，大漆，螺钿

Artist: Zhou Ziqi（China）
Type: other
Material: silver, parts, pulp,
lacquer, inlay

香火龙舞
INCENSE DRAGON DANCE

作者姓名： 朱欢（中国）
作品类型： 胸针
作品材质： 金色南洋珠，钛，9K 金，红宝石，钻石

Artist: Zhu Huan（China）
Type: brooch
Material: South Sea golden pearl,
titanium, 9K gold, ruby, diamond

砺行 / Honing

2021 北京国际
首饰艺术展
2021 BEIJING
INTERNATIONAL
JEWELLERY ART
EXHIBITION

259

牢笼
CAGE

作者姓名: 朱钧酉（中国）
作品类型: 项链，戒指，胸针
作品材质: 925 银，树脂

Artist: Zhu Junyou（China）
Type: necklace, ring, brooch
Material: 925 silver, resin

夏天
SUMMER

作者姓名: 朱玉（中国）
作品类型: 耳饰
作品材质: 银镀金，玛瑙

Artist: Zhu Yu（China）
Type: earring
Material: gold plated silver, agate

砺行 / Honing

2021 北京国际
首饰艺术展
2021 BEIJING
INTERNATIONAL
JEWELLERY ART
EXHIBITION

260

人的存在形态，茧
FORM OF BEING, COCOON

作者姓名： 訾梦楠（中国）
作品类型： 项链，耳饰
作品材质： PVC，线，银，亮布

Artist: Zi Mengnan（China）
Type: necklace, earring
Material: PVC, thread, silver, bright cloth

砺行 / Honing

2021 北京国际
首饰艺术展
2021 BEIJING
INTERNATIONAL
JEWELLERY ART
EXHIBITION

261

彩盒子
COLORED BOX

作者姓名:　庄冬冬（中国）
作品类型:　胸针
作品材质:　铜 , 大漆

Artist:　Zhuang Dongdong（China）
Type:　brooch
Material:　copper, lacquer

砺行 / Honing

2021 北京国际
首饰艺术展
2021 BEIJING
INTERNATIONAL
JEWELLERY ART
EXHIBITION

262

BRAND ZONE

品牌专区

无题
UNTITLED

作者姓名：　NK 东西宫（中国）
作品类型：　项链，戒指，手链，耳饰
作品材质：　贝雕，18K 黄金，珍珠，尼龙

Artist:　　NEWWWLOOK（China）
Type:　　　necklace, ring, bracelet, earring
Material:　carved shell, 18K yellow gold, pearl, nylon

砺行 / Honing

2021 北京国际
首饰艺术展
2021 BEIJING
INTERNATIONAL
JEWELLERY ART
EXHIBITION

266

砺行 / Honing

2021 北京国际
首饰艺术展

2021 BEIJING
INTERNATIONAL
JEWELLERY ART
EXHIBITION

人
HUMAN

作者姓名:　　TTF 高级珠宝 × 安尚秀（中国）
作品类型:　　耳饰，胸针
作品材质:　　925 银

Artist:　　TTF HAUTE JOAILLERIE × An Shangxiu（China）
Type:　　earring, brooch
Material:　　925 silver

砺行 / Honing

2021 北京国际
首饰艺术展
2021 BEIJING
INTERNATIONAL
JEWELLERY ART
EXHIBITION

268

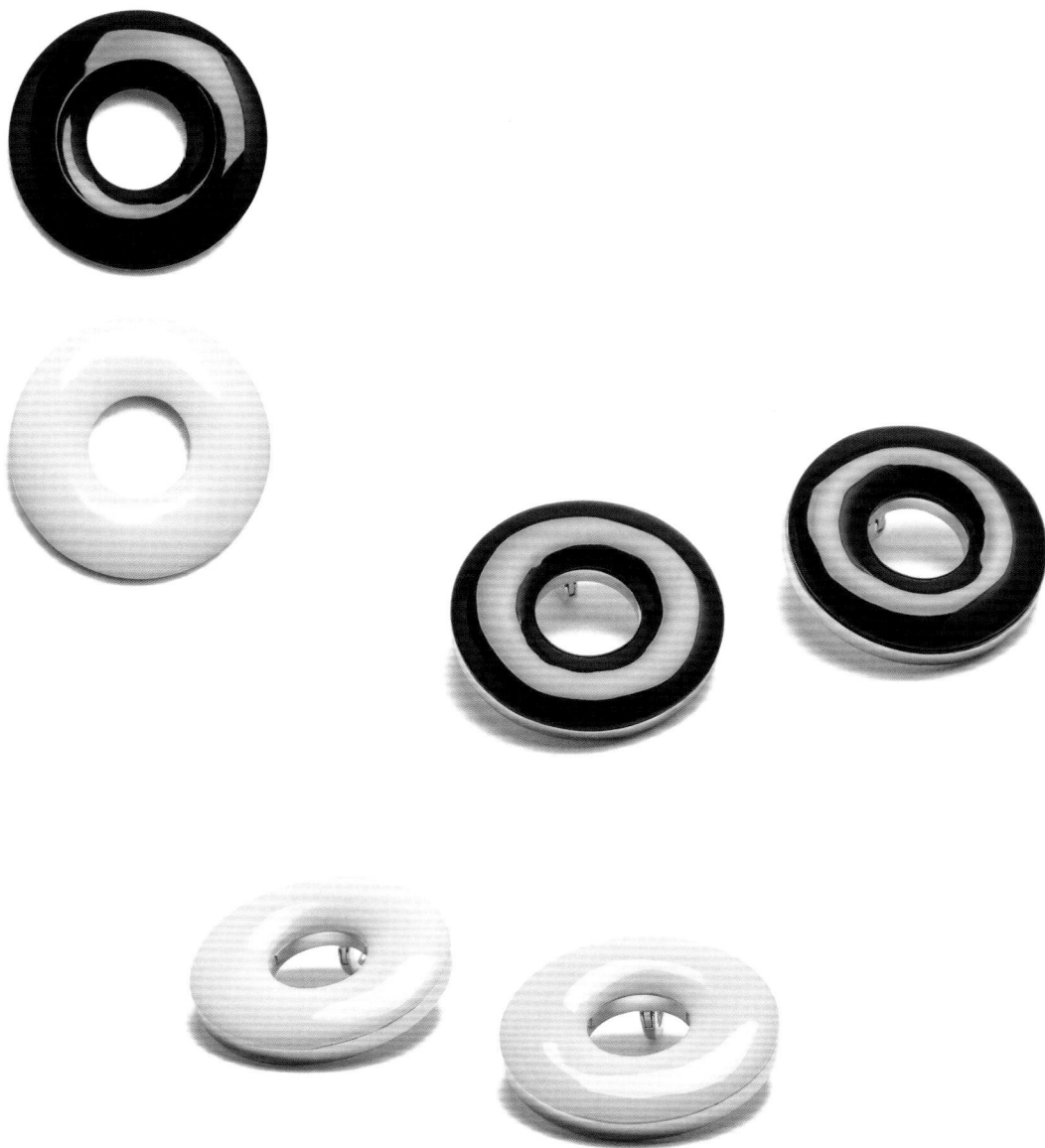

2021 北京国际
首饰艺术展
2021 BEIJING
INTERNATIONAL
JEWELLERY ART
EXHIBITION

珠联璧合系列
PERFECT PAIR SERIES

作者姓名: 菜百首饰（中国）
作品类型: 手镯，项链，耳饰
作品材质: 黄金，水晶，鱼线

Artist: Cai Bai Jewellery（China）
Type: bracelet, necklace, earring
Material: gold, crystal, fishing line

砺行 / Honing

2021 北京国际
首饰艺术展
2021 BEIJING
INTERNATIONAL
JEWELLERY ART
EXHIBITION

270

2021 北京国际
首饰艺术展
2021 BEIJING
INTERNATIONAL
JEWELLERY ART
EXHIBITION

271

大方元细金工艺品
DA FANG YUAN FINE GOLD ARTWARE

作者姓名： 北京大方元珠宝行（中国）
作品类型： 手镯，项链，耳饰，胸针，戒指，摆件
作品材质： 白水晶，珐琅，玛瑙，金，绿松

Artist:　　 Beijing Dafangyuan Jewellery Industry（China）
Type:　　　 bracelet, necklace, earring, brooch, ring, decoration
Material:　 crystal, enamel, agate, gold, kallaite

砺行 / Honing

2021 北京国际
首饰艺术展
2021 BEIJING
INTERNATIONAL
JEWELLERY ART
EXHIBITION

272

2021 北京国际
首饰艺术展
2021 BEIJING
INTERNATIONAL
JEWELLERY ART
EXHIBITION

征服王的荣耀，伊甸园，精卫填海，
飞天，遇见战国，神秘剧场
THE GLORY OF CONQUERING THE KING,
THE GARDEN OF EDEN, NEVER YIELD IN SPITE OF REVERSE,
FLYING, MEET THE WARRING STATES, MYSTERIOUS THEATRE

作者姓名: 北京正德东奇珠宝有限责任公司（中国）
作品类型: 项链，摆件
作品材质: 18K 黄金，欧泊，钻石，沙弗莱，
摩根石，彩色蓝宝石，珊瑚，红宝石，
黄色蓝宝石，托帕石，黄水晶

Artist: Beijing ZhengDeDongQi Jewellery Co., Ltd（China）

Type: necklace, decoration

Material: 18K yellow gold, opal, diamond, tsavorite,
morganite, sapphire, coral, ruby,
yellow sapphire, topaz, yellow crystal

砺行 / Honing

2021 北京国际
首饰艺术展
2021 BEIJING
INTERNATIONAL
JEWELLERY ART
EXHIBITION

274

2021 北京国际
首饰艺术展
2021 BEIJING
INTERNATIONAL
JEWELLERY ART
EXHIBITION

275

无题
UNTITLED

作者姓名： 广州市番景文化交流有限公司（中国）
作品类型： 手镯，项链，耳饰，戒指
作品材质： 钛金属，珐琅

Artist: Guangzhou Fanjing Culture Exchange Co.,Ltd（China）
Type: bracelet, necklace, earring, ring
Material: titanium, enamel

砺行 / Honing

2021 北京国际
首饰艺术展
2021 BEIJING
INTERNATIONAL
JEWELLERY ART
EXHIBITION

276

2021 北京国际
首饰艺术展
2021 BEIJING
INTERNATIONAL
JEWELLERY ART
EXHIBITION

277

梦蝶轩珐琅首饰套件
ENAMEL JEWELLERY SET BY MENGDIEXUAN

作者姓名:　上海老凤祥珐琅艺术有限公司（中国）
作品类型:　项链，耳饰，胸针
作品材质:　珐琅

Artist:　　Shanghai Laofengxiang Enamel Art Co., Ltd （China）
Type:　　　necklace, earring, brooch
Material:　enamel

砺行 / Honing

2021 北京国际
首饰艺术展
2021 BEIJING
INTERNATIONAL
JEWELLERY ART
EXHIBITION

278

2021 北京国际
首饰艺术展
2021 BEIJING
INTERNATIONAL
JEWELLERY ART
EXHIBITION

279

竹韵年年
BAMBOO FORM YEAR TO YEAR

作者姓名： 上海老庙黄金有限公司（中国）
作品类型： 项链，耳饰
作品材质： 18K 黄金

Artist: Shanghai Laomiao Jewellery Co., Ltd（China）
Type: necklace, earring
Material: 18K yellow gold

砺行 / Honing

2021 北京国际
首饰艺术展
2021 BEIJING
INTERNATIONAL
JEWELLERY ART
EXHIBITION

280

2021 北京国际
首饰艺术展
2021 BEIJING
INTERNATIONAL
JEWELLERY ART
EXHIBITION

作者姓名： 深圳市盛峰黄金有限公司（中国）
作品类型： 项链，耳饰
作品材质： 古法金，珐琅

Artist: Shenzhen Shengfeng Gold Co., Ltd（China）
Type: necklace, earring
Material: gold, enamel

砺行 / Honing

2021 北京国际
首饰艺术展
2021 BEIJING
INTERNATIONAL
JEWELLERY ART
EXHIBITION

282

2021 北京国际
首饰艺术展
2021 BEIJING
INTERNATIONAL
JEWELLERY ART
EXHIBITION

梵高
VINCENT VAN GOGH

作者姓名： 周大生珠宝股份有限公司（中国）
作品类型： 手镯，项链，耳饰，戒指，胸针
作品材质： 18K 黄金，钻石

Artist: CHOW TAI SENG Jewellery Co., Ltd（China）
Type: bracelet, necklace, earring, ring, brooch
Material: 18K yellow gold, diamond

砺行 / Honing

2021 北京国际
首饰艺术展
2021 BEIJING
INTERNATIONAL
JEWELLERY ART
EXHIBITION

284

师行 / Honing

2021 北京国际
首饰艺术展
2021 BEIJING
INTERNATIONAL
JEWELLERY ART
EXHIBITION

285